Simple Models for Understanding Nuclear Reactor Physics

Bertrand Mercier

Simple Models for Understanding Nuclear Reactor Physics

Springer

Bertrand Mercier
INSTN
Commissariat à l'Energie Atomique et aux
Energies Alternatives
Saclay, France

ISBN 978-3-031-98531-7 ISBN 978-3-031-98532-4 (eBook)
https://doi.org/10.1007/978-3-031-98532-4

This Springer imprint is published by the registered company Springer Nature Switzerland AG
The registered company address is: Gewerbestrasse 11, 6330 Cham, Switzerland

Preface

There are many excellent books dedicated to Neutron Physics and Nuclear Reactor Physics, so why write another book?

I have been teaching the subject for about 10 years, in Conservatoire National des Arts et Métiers (Cnam) and then in Ecole Nationale Supérieure des Arts et Métiers (ENSAM), so the present book contains a feedback from what the students told me.

Originally, I am an applied mathematician, and, of course, my teaching is oriented by what I am.

I like well posed models, and rigorous proofs, but students don't like them too much, so some of them will be found in annex.

My experience, is that students may understand the ideas given in the lectures if they solve exercises in Tutorials. This is why there are many exercises in this book.

Solutions are given in the appendix, so that the readers may find an incentive to solve exercises by themselves.

The models given in this book are supposed to be simple, insofar as analytical solutions can be found at least for Chaps. 1–15.

Chapter 16 is based on a joint work with Bruno Tarride (CEA/INSTN).

Chapter 17 results from my CEA activity at the Sino-French Institute for Nuclear Energy and Technology of Sun Yat-sen University, in China. There, I was lucky enough to supervise some excellent students preparing a bachelor thesis and not afraid by mathematics.

In Chaps. 16 and 17 the models have no analytical solutions.

1D finite difference models that can be easily implemented with PYTHON (enhanced by NUMPY), can solve them with a good approximation.

My experience is that engineers, who may have access to high performance computing and advanced scientific software, also need simple models to be convinced that the results they produce are reliable, and to understand what is behind.

Paris, France
December 2024

Bertrand Mercier

Acknowledgements

First, I would like to thank Laurent Turpin for welcoming me at CEA/INSTN in 2012.

INSTN was really a stimulating environment with colleagues like Paul Reuss, Jean-Christophe Klein, Bruno Tarride, Hubert Grard, Eric Royer, Hervé Golfier, Mathilde Moiron, Claude Renault and many others.

Paul Reuss is the author of *Neutron Physics* (700 pages) that is some kind of a bible in the domain, and we had very fruitful discussions. After his retirement we continued to exchange by mail, and this is how we published a joint paper (ref [14]).

I also had fruitful exchanges with Jean-Christophe Klein who teaches, like me, at the Sino-French Institute for Nuclear Energy and Technology of Sun Yat-sen University, in China.

At the same period I had the honor of being selected at Cnam as a professor "titulaire de chaire industrielle" and this is how I met Jean-Louis Nigon to whom I borrowed some ideas and some Figures.

Another bible is the book *The Physics of Nuclear Reactors* written by Serge Marguet, when he was at EDF/R&D. I never met him but I read, with much interest, many chapters of his 1200 pages book.

Many years before I arrived at INSTN, I contributed to the famous book *Mathematical Analysis and Numerical Methods for Science and Technology* by Robert Dautray and Jacques-Louis Lions.

They had both bright ideas, and I am indebted to both of them.

Jacques-Louis Lions (R.I.P.) has been my director and my mentor throughout my thesis, and even after, when I was in Aerospatiale. An outstanding mathematician.

Robert Dautray (R.I.P.) has also been my director and my mentor during my eight years in CEA/DAM: I will never forget he told me that a non-linear problem may be well posed while the associated linearized problem is ill-posed. We have beautiful examples in Chap. 15 and Sect. 17.5.

I would like also to thank my wife, Nathalie, and my son, Thibaud, for their precious support.

Bertrand Mercier

Contents

Chapter 1
Chain Reaction Principle and Effective Multiplication Factor [1–3]

In the reactor core, there is a mixture of fissile nuclei and sterile nuclei capable of absorbing neutrons. Let us recall that absorption = capture + fission; assuming that each fission gives rise to an average of 2.5 neutrons, a chain reaction may occur.

Let's suppose that in the first generation there are 1000 fissions in the core. These 1000 fissions will give rise to the creation of 2500 neutrons. Given that we can adjust the proportion between leaks, captures and fissions in the next generation, we can obtain 1000 k fissions, where k is the effective multiplication factor.

- If those 2500 lead to 50 leaks, 1450 captures and 1000 fissions, then $k = 1$
- If they lead to 50 leaks, 1449 captures and 1001 fissions then $k > 1$
- If they lead to 50 leaks, 1452 captures and 998 fissions then $k < 1$.

From generation to generation, one will obtain 1000 k^p fissions at the pth generation, in such a way that the number of fissions will increase exponentially if $k > 1$ (super-critical case). On the contrary it will stop if $k < 1$ (sub-critical): k is called the *effective multiplication factor*.

A careful control of the reactor power, and then of k, is needed. This will be explained later.

It is useful to introduce also the infinite multiplication factor k_∞ that we obtain when we neglect leaks. Obviously, $k = k_\infty \cdot$ leakage probability, so $k_\infty > 1$ is a necessary (but not sufficient) condition for the reactor to work.

B. Mercier, *Simple Models for Understanding Nuclear Reactor Physics*,
https://doi.org/10.1007/978-3-031-98532-4_1

Chapter 2
Reminders

In this chapter, we recall what is the expected value of a real random variable, and the difference between microscopic and macroscopic cross sections.

2.1 Real Random Variables

1. *Real random variables.*

A real random variable X is defined by a continuous probability law such that, if Proba of an interval set denotes the probability of the real random variable to belong to this interval:

$$\text{Proba}\{x \leq X \leq x + \mathrm{d}x\} = p(x)\mathrm{d}x$$

A distribution function can be associated in the following way:

$$F(x) = \int_{-\infty}^{x} p(y)\mathrm{d}y$$

F is the primitive of the real variable function $x \rightarrow p(x)$.
For p to be a probability distribution, it is necessary that

$$F(\infty) = \int_{-\infty}^{\infty} p(y)\mathrm{d}y = 1$$

Obviously, we have

B. Mercier, *Simple Models for Understanding Nuclear Reactor Physics*,
https://doi.org/10.1007/978-3-031-98532-4_2

$$\text{Proba}\{a \leq X \leq b\} = F(b) - F(a) = \int_a^b p(y)\mathrm{d}y$$

2. *Expected value.*

It is defined as

$$E[X] = \int_{-\infty}^{\infty} x p(x)\mathrm{d}x$$

It is then a weighted average of x.

Examples

$$p(x) = \left\{ \begin{matrix} \frac{1}{b-a} \text{ if } a \leq x \leq b \\ 0 \quad \text{otherwise} \end{matrix} \right\} \text{ then } E[X] = \frac{1}{b-a}\int_a^b x\mathrm{d}x = \frac{a+b}{2}$$

$$p(x) = \left\{ \begin{matrix} 2x \text{ if } 0 \leq x \leq 1 \\ 0 \quad \text{otherwise} \end{matrix} \right\} \text{ then } E[X] = 2\int_0^1 x^2\mathrm{d}x = \frac{2}{3} > \frac{1}{2}$$

$$p(x) = \left\{ \begin{matrix} 2(1-x) \text{ if } 0 \leq x \leq 1 \\ 0 \quad \text{otherwise} \end{matrix} \right\} \text{ then } E[X] = 2\int_0^1 x(1-x)\mathrm{d}x = \frac{1}{3} < \frac{1}{2}$$

3. *Example of a real random variable useful in neutronics.*

$$p(x) = \left\{ \begin{matrix} \sigma \mathrm{e}^{-\sigma x} \text{ if } x \geq 0 \\ 0 \quad \text{otherwise} \end{matrix} \right\}$$

We check that

$$F(x) = \int_0^x \sigma \mathrm{e}^{-\sigma y}\mathrm{d}y = 1 - \mathrm{e}^{-\sigma x}$$

and

$$F(\infty) = 1$$

Expected value:

$$E[X] = \int_0^\infty x\sigma e^{-\sigma x} dx$$

We notice that $\frac{d}{dx} xe^{-\sigma x} = e^{-\sigma x} - x\sigma e^{-\sigma x}$ and then (by integration from 0 to ∞):

$$E[X] = \frac{1}{\sigma} F(\infty) = \frac{1}{\sigma}$$

■

2.2 Cross Sections [1]

Just because a reaction is possible doesn't mean it happens. In this paragraph, we shall focus on neutrons.

Microscopic Cross Section

It is equal to the area of some circle centered around the nucleus of the target atom, s.t.:

- either the neutron passes through the circle and it interacts with the atom,
- or it does not, and it does not interact.

Such a surface area is usually very small; it is measured in *barn (b)*

$$1b = 10^{-24}\,\text{cm}^2.$$

This is equivalent to modeling the nucleus as a hard sphere.

The physical reality is more complex, as particles are actually waves.

The interaction between two waves can occur even at distances not directly related to the diameter of the nucleus.

This explains why, for two types of interaction, the microscopic effective cross-section can be different. This is the case for the hydrogen atom, for example,

$$\sigma_{\text{coll}} = 20b \text{ for elastic collisions}$$

$$\sigma_{\text{capt}} = 0.322b \text{ for capture}$$

Neutron Flux

We shall define the neutron current later (Chap. 8).

Monokinetic case: all neutrons travel at the same velocity v. We assume a steady state flux.

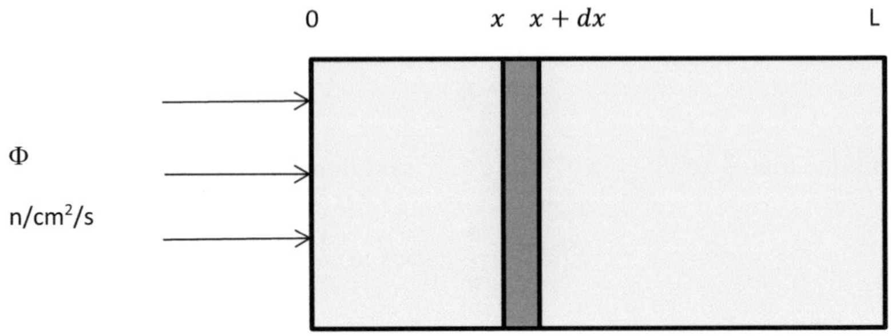

Fig. 2.1 Slab of material

If $n = n(x, t)$ denotes the number of neutrons per cm^3, then $\Phi = nv$; v being measured in cm/s, $\Phi = nv$ is then measured in n/cm^2/s.

Macroscopic Cross Section

We assume an homogeneous medium containing N_o at/cm^3 of a given nucleus.

We consider an incident beam with neutrons all moving in the same direction and at the same speed v.

We consider a slice of material within the strip $(0, L)$ (see Fig. 2.1).

In any cylinder with section 1 cm^2 and thickness dx (and then with volume dx cm^3) there are $N_o \cdot \mathrm{d}x$ nuclei.

If dx is small enough, there are no two nuclei one behind the other in the direction of the x-axis.

The obstacle formed by the $N_o \cdot \mathrm{d}x$ nuclei has surface area $N_o\sigma \cdot \mathrm{d}x$ if σ denotes the microscopic cross-section of a nucleus. The probability of a neutron interacting with an atom is therefore $N_o\sigma \cdot \mathrm{d}x$. Thus, the flux Φ will decrease by

$$-d\Phi = \Phi N_o\sigma \cdot \mathrm{d}x.$$

Let $\Sigma = N_o\sigma$, the neutron flux $\Phi = \Phi(x)$ is then solution of the ordinary differential equation

$$\frac{\mathrm{d}}{\mathrm{d}x}\Phi = -\Sigma\Phi$$

hence

$$\Phi(x) = \Phi(0)\mathrm{e}^{-\Sigma x};$$

Σ measured in cm^{-1} is the macroscopic cross section.

Note: The number of interactions per cm^3 per second is equal to $\Sigma\Phi$. It is called *reaction rate*.

Let's assume that a reactor core is homogeneous and infinite (no leakage).

Let us denote by Σ_f (resp. Σ_c) the fission (resp. capture) macroscopic cross section, the consequence of absorption = capture + fission is then that

$$\Sigma_a = \Sigma_f + \Sigma_c$$

so that in each cm^3 of the core, there is $\Sigma_f\Phi$ fissions and $\Sigma_a\Phi$ absorptions per second.

Let ν denote the average number of neutrons generated by each fission (we have $\nu = 2.42$ for ^{235}U and $\nu = 2.91$ for ^{239}Pu) then the effective multiplication factor will be

$$k_\infty = \nu \frac{\Sigma_f \Phi}{\Sigma_a \Phi} = \nu \frac{\Sigma_f}{\Sigma_a} \tag{2.1}$$

Note that in an infinite homogeneous reactor, a local analysis is sufficient:

- all neutrons are eventually absorbed
- $\frac{\Sigma_f}{\Sigma_a}$ is the fission probability
- $\frac{\Sigma_c}{\Sigma_a}$ is the capture probability
- $\frac{\Sigma_f}{\Sigma_a} + \frac{\Sigma_c}{\Sigma_a} = 1.$

It's true that neutrons may have elastic collisions that change their direction, but these changes in direction do not matter in an infinite homogeneous reactor.

In the case of a finite reactor, even a homogeneous one, *leakage* will have to be taken into account, and the effective multiplication coefficient k will satisfy

$$k = \nu \frac{\text{fissions}}{\text{absorptions} + \text{leakage}}$$

And the probability of fission will be equal to the ratio $\frac{\text{fissions}}{\text{absorptions}+\text{leakage}}$.

If $\nu = 2.5$, it will be necessary to tune capture in such a way that this probability be equal to 40%.

Exercise 2.1: PWR In a PWR 1300 the fuel volume is about 11.1 m^3: the core power is about 3800 MWth. Given that the energy released by a ^{235}U fission is equal to 200 MeV, compute the number of fissions per fuel cm^3 per second.

Given that 83% fissions are due to thermal neutrons, and that the fuel fission macroscopic cross section is $\Sigma_f \sim 0.24\,\text{cm}^{-1}$, compute the thermal neutron flux in the fuel (*solution p. 165*). ■

Chapter 3
Principle of Fast Neutron Reactors (FNR) [4, 5]

In this chapter, we show that, assuming an homogeneous core, the infinite multiplication factor k_∞ is quite easy to evaluate for fast neutron reactors.

The fission neutron spectrum has the shape appearing on Fig. 3.1.

The analytical formula given by Watt (see Wikipedia) is the following (cf. [2])

$$p(E) = \frac{2\exp(-AB/4)}{\sqrt{\pi A^3 B}} \exp(-E/A) \mathrm{sh} \sqrt{BE}$$

with $A = 0.965\,\mathrm{MeV}$ and $B = 2.29\,\mathrm{MeV}^{-1}$ for ^{235}U.

Fast neutron reactors usually consume MOX fuel, that is mixture of PuO_2 and UO_2. They all have a coolant that allows them to evacuate thermal power, but they don't have a good moderator.

In the case of sodium, that is an excellent coolant, $A = 23$, it's a very poor moderator (see Chap. 7): neutrons are slowed down very little at each collision, so they're bound to be captured by the ^{238}U traps that are defined as the peak values of its microscopic capture cross sections (see Fig. 4.2).

We can therefore consider that in a sodium reactor there are virtually only fast neutrons.

On Fig. 3.2, we show a typical example of the basic core cell of a Sodium reactor [4].

It can be easily seen that with such a pattern, 85.4% of the core is occupied by fuel pins and 14.6% by sodium. Now given the cladding thickness (0.5 mm), let V_f be the MOX volume, V_c the cladding volume, and V_N the sodium volume, we let

$$\omega_1 = V_\mathrm{f}/(V_\mathrm{f} + V_\mathrm{c} + V_\mathrm{N}) \; \omega_2 = V_\mathrm{c}/(V_\mathrm{f} + V_\mathrm{c} + V_\mathrm{N}) \; \omega_3 = V_\mathrm{f}/(V_\mathrm{f} + V_\mathrm{c} + V_\mathrm{N})$$

we easily check that $\omega_1 = 0.6515$, $\omega_2 = 0.2025$, $\omega_3 = 0.1460$.

For the sake of simplicity we assume that the cladding is purely made of ^{56}Fe.

B. Mercier, *Simple Models for Understanding Nuclear Reactor Physics*,
https://doi.org/10.1007/978-3-031-98532-4_3

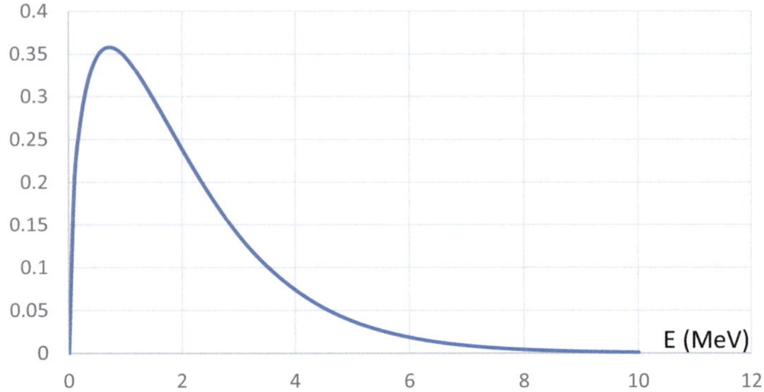

Fig. 3.1 Watt fission neutron spectrum for ^{235}U

Fig. 3.2 Unit cell in a sodium fast reactor

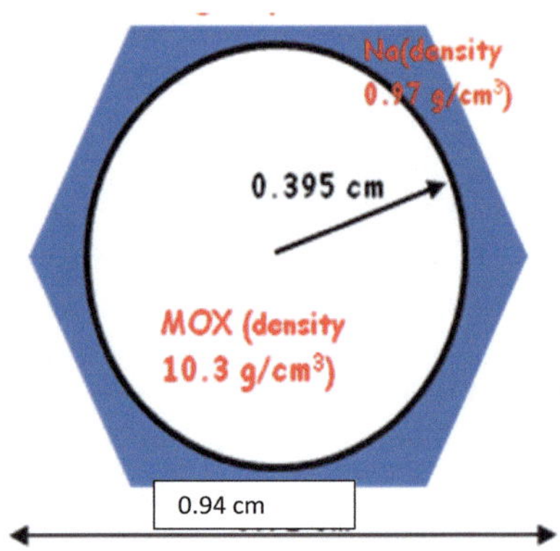

Exercise 3.1 For fast neutrons, we assume the following microscopic average cross-sections (in barn)

	Fission	Capture
^{239}Pu	1.9	0.6
^{238}U	0.04	0.3
^{56}Fe		0.1
^{23}Na		0.01

The effective oxygen capture cross-section is neglected. The velocity of all neutrons is assumed to be the same. We note that ^{238}U is slightly fissile in the fast spectrum, but not at all in the thermal spectrum. We select $\nu = 2.91$. Assuming the core is homogeneous, and applying formula (2.1) calculate k_∞ in the case where the MOX enrichment $e = 11\%$. Show that $k_\infty < 1$ if $e = 10\%$. What can you conclude? (*solution p. 165*) ■

Usually fast neutron reactor operators manufacture the MOX fuel by using Plutonium coming from reprocessing of used UO_2 fuel (see Chap. 13). Such a Plutonium typically contains 70% fissile isotopes only (59% ^{239}Pu and 11% ^{241}Pu) see https://www.global-chance.org/IMG/pdf/gc34p12-13.pdf.

Chapter 4
Why Does It Make Sense to Slow Down Neutrons?

In this chapter, we explain why neutron slowing down is useful to reduce the size of the core and to be able to use fuel with low enriched uranium.

There are international databases such as

http://www.oecd-nea.org/janisweb/index.html

which give values of microscopic cross-sections.

For example, Fig. 4.1 shows the microscopic fission cross-section of ^{235}U (expressed in barn) as a function of incident neutron energy (expressed in MeV).

It is clear that the ^{235}U fission cross-section is much stronger for thermal neutrons (whose energy is of the order of 0.025 eV) than for fast neutrons (whose energy is of the order of 1 MeV).

Since all neutrons produced by fission are fast, a natural idea is to slow them down: this is the role of the moderator.

With a moderator, smaller but equally powerful reactors can be built.

Close examination of Fig. 4.2 shows that, while ^{238}U is relatively transparent to thermal neutrons, the same cannot be said for neutrons in the epithermal range, intermediate between the fast and thermal ranges. A fast neutron remaining in the Uranium during its deceleration has every chance of being captured by the "traps" of ^{238}U (i.e. the peaks of its capture cross section). It is more likely to survive if there is sufficient moderator.

What is a Moderator?

It's a medium that slows neutrons by elastic collision without capturing them (too much).

Figure 4.3, taken from [2], shows what happens in the core of a PWR.

What we want in a PWR is for a fast neutron produced by fission to make as much of a random walk in the light water as possible, and to return to the fuel only once it is thermal, in which case, with a high probability, it will fission a ^{235}U nucleus.

B. Mercier, *Simple Models for Understanding Nuclear Reactor Physics*,
https://doi.org/10.1007/978-3-031-98532-4_4

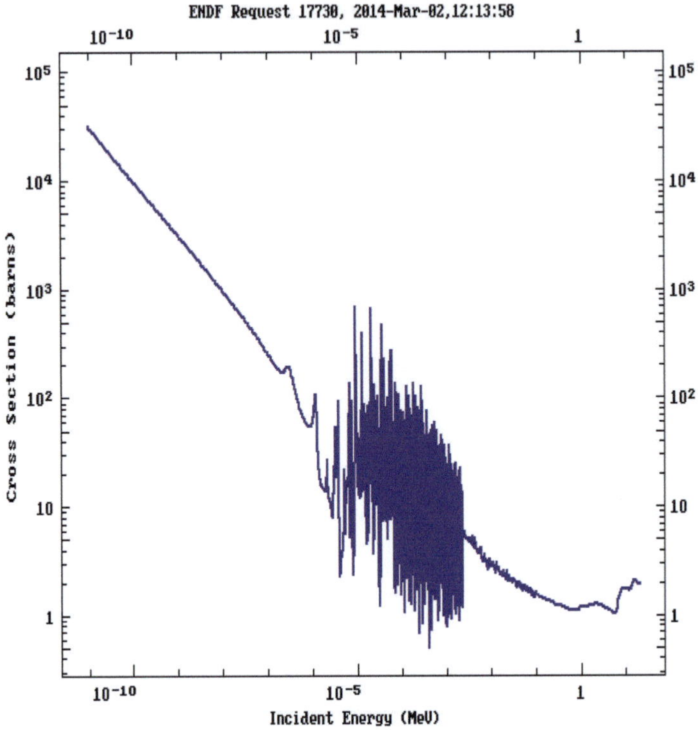

Fig. 4.1 Microscopic fission cross section of ^{235}U

Possibly, neutrons will return to fuel (represented by the 4 circles in Fig. 4.3) prematurely, and risk being captured by ^{238}U.

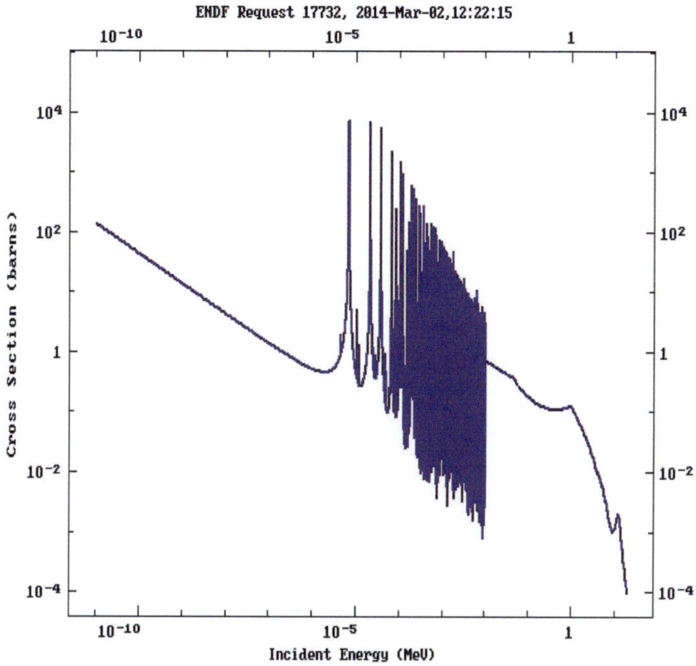

Fig. 4.2 Microscopic capture cross section of ^{238}U

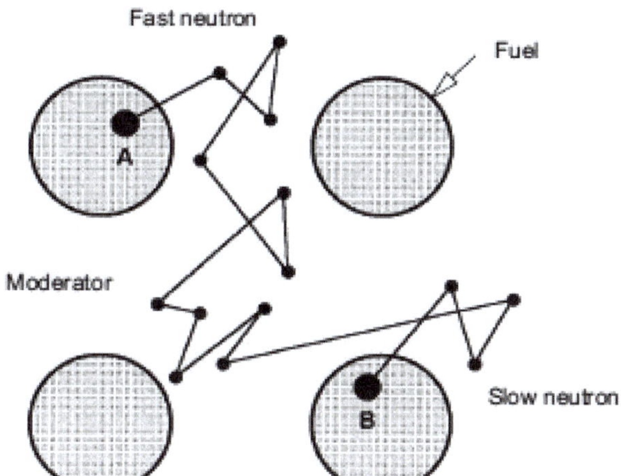

Fig. 4.3 Random walk of a neutron in the core of a PWR. From [2]

Chapter 5
Fermi's 4-Factors Formula [2, 3]

Fermi led the team that designed and built the first nuclear reactor, Chicago Pile 1 that reached criticality on December 2, 1942. On this occasion, he developed the famous 4 factors formula that helps understanding how a thermal reactor works.

In thermal neutron reactors, the aim is to take advantage of the very large microscopic fission cross-section of ^{235}U (or the even larger fission cross-section of ^{239}Pu) for thermal neutrons.

Neutrons must therefore be slowed down and thermalized, without being captured by ^{238}U traps.

We need to evaluate the probability that a fast neutron survives the slowing-down phase without being captured by ^{238}U. This is known as the *resonance escape probability*.

If the reactor could be considered infinite, we can see that between the birth of a neutron by fission and its death by capture or fission, it can happen that

- the neutron undergoes fast fission
- it is captured by ^{238}U traps,
- the neutron is slowed down and thermalized, but captured by the moderator
- this neutron, once thermalized, returns to the fuel and is absorbed, which, with a good probability, will produce a new fission.

Having made this analysis, Enrico Fermi decomposed the «infinite» multiplication factor k_∞ as the product of 4 factors:

$$k_\infty = \varepsilon\,.\,p\,.\,f\,.\,\eta \tag{5.1}$$

where

- $\varepsilon > 1$ is the *fast fission factor*: if, at generation n, there are 1000 neutrons following fast fissions this number becomes $1000\,.\,\varepsilon$; ε takes into account ^{238}U fissions by fast neutrons.

B. Mercier, *Simple Models for Understanding Nuclear Reactor Physics*,
https://doi.org/10.1007/978-3-031-98532-4_5

- $p < 1$ is the *resonance escape probability*. This is the probability for a fast neutron to be slowed down to the thermal range without being captured by the ^{238}U traps.
- $f < 1$ is the *thermal utilization factor*, i.e. the probability for a thermal neutron not to be captured by the moderator or by a structural material, and therefore of returning to the fuel.
- $\eta > 1$ is the *reproduction factor*: it is the average number of fast neutrons generated by a thermal neutron returning to the fuel: this number would be equal to ν if all the thermal neutrons absorbed in the fuel produced fissions, but η also takes into account the risk of capture of thermal neutrons by a Uranium nucleus. (^{235}U or ^{238}U).

Leakage, i.e., neutrons escaping from the core, can also be taken into account. This is shown in Fig. 5.1, taken from Ref. [6].
The effective multiplication factor k_{eff} can be written as the product of 6 factors

$$k_{\text{eff}} = \varepsilon \,.\, \mathcal{L}_f \,.\, p.\mathcal{L}_t \,.\, f \,.\, \eta \tag{5.2}$$

The two new factors that appear in this formula are

$\mathcal{L}_f < 1$ fast non-leakage probability
$\mathcal{L}_t < 1$ thermal non-leakage probability.

The difference between k_{eff} and k_∞ is that the former takes leakage into account, while the latter does not.

We always have $k_{\text{eff}} < k_\infty$.
What we want in normal operation is for the reactor to be just critical.

$$k_{\text{eff}} = 1$$

To achieve this goal, it is necessary that $k_\infty > 1$.
Otherwise, the reactor will remain subcritical and unable to operate.

How to Evaluate the 4 Fermi Factors?

The f and η factors are relatively easy to evaluate.
Let us start with f.

Assume that the core is homogeneous, i.e. that in each cm^3 of the core there is ω cm^3 of fuel and $(1-\omega)$ cm^3 of moderator ; if Φ is the thermal neutron flux in the core (defined as $\Phi = n.v$ where n is the thermal neutron density and v the velocity of thermal neutrons)
Σ_{ac} the macroscopic absorption cross section (= capture + fission) of the fuel (thermal neutrons)
Σ_{am} the macroscopic absorption cross-section of the moderator (also for thermal neutrons)
we know that every second there are $\Sigma_{ac}\Phi$ (neutron) absorptions in the fuel and $\Sigma_{am}\Phi$ absorptions in the moderator.

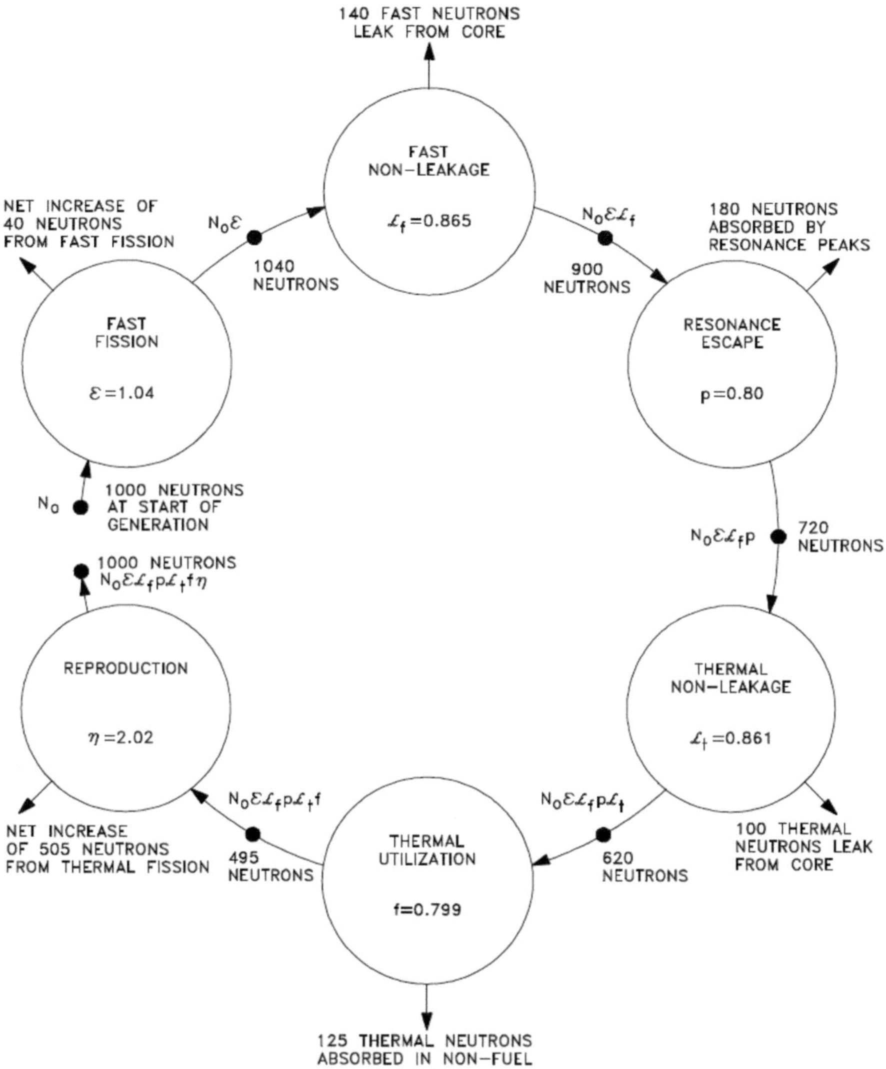

Fig. 5.1 Neutron life cycle in the core of a PWR (from [6])

Logically enough, the ratio $(1 - \omega)/\omega$ is called the moderation ratio.

Absorptions in the moderator are "sterile", unlike absorptions in the fuel, which can be fissions.

Therefore, assuming that the neutron flux is uniform, and neglecting the fuel cladding, we have

$$f = \frac{\Sigma_{ac}\Phi}{\Sigma_{ac}\Phi + \Sigma_{am}\Phi} = \frac{\Sigma_{ac}}{\Sigma_{ac} + \Sigma_{am}} \tag{5.3}$$

Exercise 5.1 (*Boron Efficiency*) At the PWR operating temperature there are 5×10^{22} at H/cm^3 in the moderator, which gives $\Sigma_{H_2O} = 0.016\,\text{cm}^{-1}$. We specify that the core contains 2/3 moderator and 1/3 fuel.

We also specify $\Sigma_{\text{fuel}} = 0.6\,\text{cm}^{-1}$, compute f.

We add 1000 ppm (parts per million) boron in water (more precisely 1 boron atom for 1000 hydrogen atoms), compute the new value of factor f given that for natural boron $\sigma_B = 600\,b$ (*solution*, p. 166). ■

As above, if Σ_{fc} denotes the macroscopic fission cross section of the fuel, the probability of fission for an absorption in the fuel will be

$$P_{\text{fis}} = \frac{\Sigma_{fc}}{\Sigma_{ac}}$$

If ν denotes the average number of neutrons produced by each fission ($\nu \cong 2.42$ for ^{235}U) we shall have

$$\eta = \nu P_{\text{fis}} = \nu \frac{\Sigma_{fc}}{\Sigma_{ac}} \tag{5.4}$$

Assuming there are N_U uranium atoms per cm^3 in the fuel, and e denotes its enrichment, there will be eN_U ^{235}U atoms and $(1-e)N_U$ ^{238}U atoms per cm^3 in the fuel.

For thermal neutrons we have $\sigma_{f5} \cong 580\,b$, $\sigma_{a5} \cong 680\,b$, $\sigma_{f8} \cong 0\,b$, $\sigma_{a8} \cong 2.7\,b$.

We deduce that

$$\eta = \nu \frac{\Sigma_{fc}}{\Sigma_{ac}} = \nu \frac{e\sigma_{f5}}{e\sigma_{a5} + (1-e)\sigma_{a8}} \tag{5.5}$$

is greater, the higher the enrichment e.

Enrico Fermi had no enriched uranium; only natural uranium for which $e \cong 0.72\%$. He could therefore rely only on $\eta \cong 1.31$.

He had to work a lot to obtain a p-factor and an f-factor as close as possible to 1.

His reactor, that used graphite as a moderator, could not work in the homogeneous case as mentioned in Chap. 7.

As we shall see in Fig 13.3, if only natural uranium is available, light water cannot be used as a moderator, as it is (somewhat) too absorbent. Instead, heavy water (D_2O) or graphite must be used. Graphite was chosen by Enrico Fermi to build the world's first reactor in Chicago in 1942, as he did not have a sufficient stock of heavy water.

As we shall see, elastic collisions slow down neutrons.

Exercice 5.2 In a PWR, we assume that over 100 neutrons having survived to slowing down, 60 make a fission, 37 are captured (18 in the fuel + 19 in the moderator) and 3 escape from the core.

We assume that those 60 fissions generate 144 fast neutrons ($\nu = 2.4$).

On these 144 fast neutrons, 3 produce (fast) fissions (then generate 9 fast neutrons).

Then, there are 150 fast neutrons entering in the slowing-down phase, 33 are captured during this phase and 17 escape from the core.

We get that 100 neutrons survive to slowing down, so the loop is complete

1. Compute the 6 factors and check that their product is equal to 1.
2. Compute k_∞.
3. Given that the fuel is uranium based, and that, for thermal neutrons (with obvious notations):

$$\sigma_f^5 = 580\,b; \quad \sigma_a^5 = 680\,b; \quad \sigma_f^8 = 0\,b; \quad \sigma_a^8 = 2.7\,b$$

compute the fuel enrichment (*solution*, p. 166). ■

Chapter 6
Neutron-Nucleus Elastic Collisions

In this chapter, we show that elastic collisions on light nuclei are producing neutron slowing down.

More generally, we assume that particle A with mass m_A and velocity $\vec{v}$ collides with a target B (assumed motionless) with mass m_B (Fig. 6.1).

After collision, particle A is assumed to have velocity $\vec{V}_A'$ and target B velocity $\vec{V}_B'$ (see Fig. 6.2).

Elastic collision: both momentum and kinetic energy are conserved.

In Annex 1 we show impulsions before and after impact in the center of mass and we prove the following formula:

$$E_A' = \frac{1}{2}E_A(1 + \gamma + (1 - \gamma)\mu) \tag{6.1}$$

where

$E_A = \frac{1}{2}m_A v^2$ is the kinetic energy *before* impact
$E_A' = \frac{1}{2}m_A\left(V_A'\right)^2$ is the kinetic energy *after* impact

$$\gamma = \frac{(x-1)^2}{(x+1)^2},\quad x = \frac{m_B}{m_A}$$

μ a random number uniformly distributed in the interval (− 1, + 1).

We easily deduce that $E_A' \in \left[\gamma E_A, E_A\right]$ and that

$$\text{expected value}\left(E_A'\right) = \frac{1+\gamma}{2}E_A$$

Light nuclei are best at slowing down neutrons: the best is undoubtedly the hydrogen atom. ($A = 1$) that leads to $\gamma = 0$; with hydrogen, on average, the neutron

B. Mercier, *Simple Models for Understanding Nuclear Reactor Physics*,
https://doi.org/10.1007/978-3-031-98532-4_6

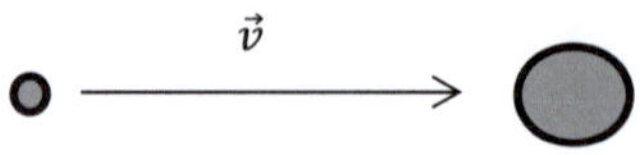

Fig. 6.1 Particle A colliding with particle B.

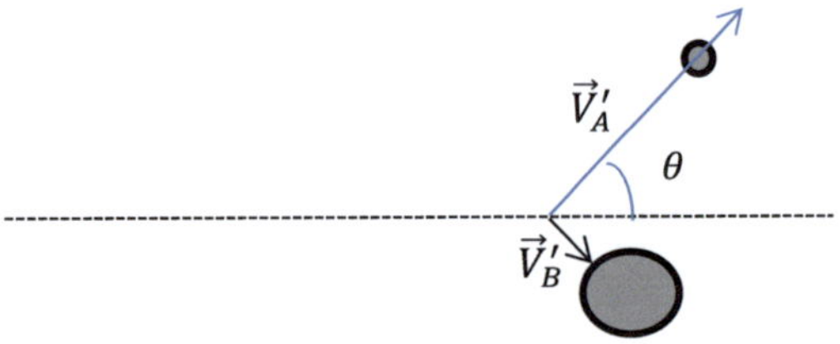

Fig. 6.2 After collision (laboratory framework)

energy is divided by 2 at each collision. So in 10 collisions the neutron energy is divided by 1024, and in 20 collisions on a hydrogen atom, a fast neutron becomes thermal.

Figure 4.3 illustrated what happens in the core of a PWR.

What we want in a PWR is for a fast neutron produced by fission to make, as far as possible, a random walk in the light water and to return to the fuel only when it is thermal, in which case, with a high probability, it will fission a ^{235}U nucleus.

Inevitably, some neutrons will return to the fuel (represented by the 4 circles in Fig. 4.3) prematurely and risk being captured by some ^{238}U nucleus.

Chapter 7
Evaluation of the Resonance Escape Probability

In this chapter, we introduce the slowing down equation to evaluate factor p from Fermi's 4 factors formula. We assume an homogeneous core.

To evaluate the resonance escape probability p, we shall proceed to a change of variable: we replace energy E by lethargy u defined by

$$u = \ln\left(\frac{E_0}{E}\right)$$

where E_0 is a reference energy often taken to be equal to 10 MeV.

As we have seen (Fig. 3.1), there are no fission neutrons of energy greater than 10 MeV, so the lethargy of all neutrons moving through the core will be positive.

Exercise 7.1 Compute E when $u = 7$; then when $u = 15$. (*solution* p. 167) ■

We saw in Chap. 6 that in the case of elastic collision, the energy E' after shock is related to the energy E before collision by the formula

$$E' = \frac{1}{2}E(1 + \gamma - (1 - \gamma)\mu) \quad \text{where} \quad \gamma = (x - 1)^2/(x + 1)^2$$

and $\mu \in [-1, +1]$ is a uniformly distributed random variable, which means that the probability for

$\mu \in [\mu_a, \mu_b]$ is equal to $\frac{\mu_b - \mu_a}{2}$ if $\mu_a < \mu_b$.

Let $u = \ln\left(\frac{E_0}{E}\right)$ and $u' = \ln\left(\frac{E_0}{E'}\right)$, the lethargic jump due to the collision is therefore

$$w = u' - u = -\ln\left(\frac{1 + \gamma - (1 - \gamma)\mu}{2}\right),$$

We note that if $\mu = -1$, $w = 0$ and if $\mu = 1$, $w = \varepsilon$ where $\varepsilon = -\ln \gamma$.

More precisely if $\mu \in [-1, +1]$ then $w \in [0, \varepsilon]$.

B. Mercier, *Simple Models for Understanding Nuclear Reactor Physics*,
https://doi.org/10.1007/978-3-031-98532-4_7

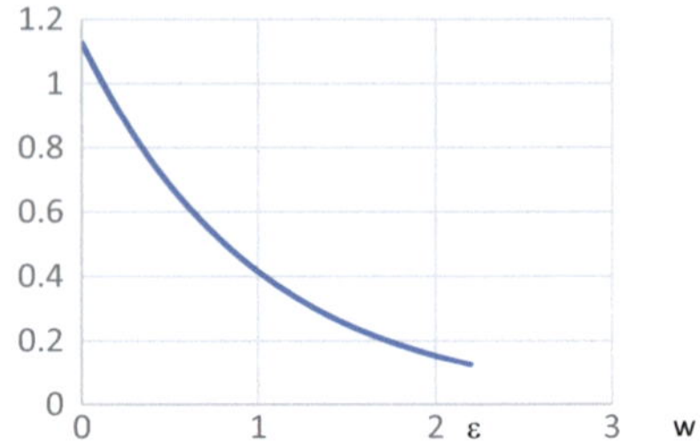

Fig. 7.1 Probability distribution of lethargy jump w (case $x = 2$ then $\varepsilon \cong 2.2$)

Since $\mu \in [-1, +1]$, $w \in \left[0, -\ln\gamma\right]$

$$\mathbb{E}\,\mathrm{w} = \frac{1}{2}\int_{-1}^{1} w\mathrm{d}\mu = -\frac{1}{2}\int_{-1}^{1} \ln\left(\frac{1}{2}((1+\gamma) - (1-\gamma)\mu)\right)\mathrm{d}\mu$$

To carry out the change of variable $\mu \to w$ in the integral, we need to know what to replace $\mathrm{d}\mu$ with.

We have

$$\frac{1}{2}(1+\gamma-(1-\gamma)\mu) = \mathrm{e}^{-w}$$

And then

$$\frac{1}{2}\mathrm{d}\mu = \frac{1}{1-\gamma}e^{-w}\mathrm{d}w$$

Hence, for the expected value

$$E[w] = \int_{0}^{-Ln\gamma} w\frac{1}{1-\gamma}\mathrm{e}^{-w}\mathrm{d}w$$

It also proves that the probability distribution of the random variable w satisfies

$$p(w) = \frac{1}{1-\gamma}\mathrm{e}^{-w} \quad \text{on interval } [0, \varepsilon] \tag{7.1}$$

(see Fig. 7.1)

Exercise 7.2 Prove that $\int_0^\varepsilon p(w)\mathrm{d}w = 1$ and that $\xi = E[w] = 1 - \frac{\gamma\varepsilon}{1-\gamma}$ ■ (*solution* p. 167)

Slowing down equation.

We have seen that if

$\Phi = \Phi(u)$ is the neutron flux (measured in neutrons/cm^2/s/lethargy unit),
$\Sigma = \Sigma(u)$ (in cm^{-1}), the macroscopic collision cross section

Then $\Sigma(u)\Phi(u)$ is the collision density. (/cm^3/s/ lethargy unit) i-e $\Sigma(u)\Phi(u)\mathrm{d}u$ is the number of events per cm^3 per second in a lethargy interval du around u.

Actually $\Sigma(u) = \Sigma_s(u) + \Sigma_a(u)$ then, in case there is a collision, our neutron disappears with probability $\Sigma_a(u)/\Sigma(u)$.

With probability $\Sigma_s(u)/\Sigma(u)$ it will be an elastic collision and the probability for the new lethargy to be u' after collision will be (see (7.1))

$$p(u, u') = \frac{1}{1-\gamma}\mathrm{e}^{u-u'} \tag{7.2}$$

(note that $u + \varepsilon > u' > u$).

Now, let us consider a neutron source $S = S(u)$ (such that $S(u)$ is the number of neutrons generated per cm^3/s/lethargy unit).

We assume that $S(u) = 0$ if $u < 0$.

In a steady state, there must be as many neutrons appearing as neutrons disappearing in the lethargy band $(u,\ u + du)$.

Then, we must have

$$S(u) + \int_{u-\varepsilon}^{u} \Sigma_s(u')\Phi(u')p(u', u)\mathrm{d}u' = \Sigma(u)\Phi(u) \tag{7.3}$$

In fact, the second term takes into account neutrons that appeared at lethargy u after having had a collision at lethargy u' $(u - \varepsilon < u' < u)$

We shall prove the following result :

Proposition 7.1 *When the moderator is hydrogen* ($\gamma = 0$ so that $\xi = 1$), *then if* p *is the resonance escape probability we have*:

$$p = \exp\left(-\int_{u_0}^{u_1} \frac{\Sigma_a(u')}{\Sigma(u')}\mathrm{d}u'\right) \tag{7.4}$$

where (u_0, u_1) *is a lethargy interval containing the* ^{238}U *traps.*

Proof Let us introduce the neutron current

$$q(u) = \int_0^u \mathrm{d}u' \int_u^{\infty} \Sigma_s(u')\Phi(u')p(u', u'')\mathrm{d}u''$$

(note that $q(u)$ counts all neutrons that had an elastic collision at lethargy $u' < u$, collision that shifted their lethargy to $u'' > u$).

The following relation can be easily understood

$$q(u + \mathrm{d}u) - q(u) = S(u)\mathrm{d}u - \Sigma_a(u)\,\Phi(u)\mathrm{d}u$$

(indeed the neutron current is increased by all neutrons emitted by source $S(u)$ and decreased by all absorptions).

We deduce:

$$\frac{d}{\mathrm{d}u}q(u) = S(u) - \Sigma_a(u)\Phi(u) \tag{7.5}$$

that is our first relation.

We shall obtain a second one as follows

Lemma 7.1 *In the case where*

$$p(u', u) = \left\{\begin{matrix} \mathrm{e}^{u'-u} & \text{if} \quad u' \le u \\ 0 & \text{otherwise} \end{matrix}\right\} \tag{7.6}$$

We have

$$\frac{\mathrm{d}q}{\mathrm{d}u} + \frac{\Sigma_a(u)}{\Sigma(u)}q = \frac{\Sigma_s(u)}{\Sigma(u)}S(u). \tag{7.7}$$

Proof:

When $p(u', u)$ is given by (7.6), we have:

$$q(u) = \int_0^u \Sigma_s(u')\Phi(u')\mathrm{e}^{u'}\mathrm{d}u' \int_u^\infty \mathrm{e}^{-u''}\mathrm{d}u'' = \int_0^u \Sigma_s(u')\Phi(u')\mathrm{e}^{u'}\mathrm{d}u'\mathrm{e}^{-u}$$

$$= \int_0^u \Sigma_s(u')\Phi(u')p(u', u)\mathrm{d}u' = \Sigma(u)\Phi(u) - S(u)$$

So that:

$$\Phi(u) = \frac{S(u)}{\Sigma(u)} + \frac{q(u)}{\Sigma(u)} \tag{7.8}$$

By using (7.8), we obtain

$$\frac{\mathrm{d}q}{\mathrm{d}u} = S(u) - \Sigma_a(u)\Phi(u) = S(u) - \Sigma_a(u)\left(\frac{S(u)}{\Sigma(u)} + \frac{q(u)}{\Sigma(u)}\right)$$

hence

$$\frac{\mathrm{d}q}{\mathrm{d}u} + \frac{\Sigma_a(u)}{\Sigma(u)}q = \frac{\Sigma(u) - \Sigma_a(u)}{\Sigma(u)}S(u) = \frac{\Sigma_s(u)}{\Sigma(u)}S(u)$$

which is the desired result. ■

Particular case: $S(u) = \delta(u - v)$, i-e S is a Dirac measure in $u = v$:

$$\frac{dq}{du} + \frac{\Sigma_a(u)}{\Sigma(u)} q = 0$$

$$q(v_+) = \frac{\Sigma_s(v)}{\Sigma(v)}$$

We deduce that

$$q(u) = q(v_+) \exp\left(- \int_v^u \frac{\Sigma_a(w)}{\Sigma(w)} dw \right). \tag{7.9}$$

Since we should have $q(u) = p \cdot q(v_+)$, we obtain (7.4), which completes the proof of Proposition 7.1. ■

Remark 7.1

1. The Watt fission spectrum in Fig. 3.1 shows that fission neutron with energy less than 9 keV (then with lethargy $u > 7$) can be neglected. Then we can assume $S(u) = 0$ when $u > 7$.

In such a case (7.8) shows that $q(u) = \Sigma(u).\Phi(u)$.
Then for $u > 7$ the neutron current is decreasing.
We wave seen on Fig. 4.2 that the curve $u \rightarrow \Sigma(u)$ may have "traps" (i-e large resonance peak due to ^{238}U). Now, since $\Sigma(u).\Phi(u)$ is decreasing, those (very) large values of $\Sigma(u)$ must be balanced by (very) small values of $\Phi(u)$.
This is a *self-shielding* effect.

2. Multigroup methods require computing suitable averages of neutron cross section on some lethargy interval $[a, b]$. Obviously, due to the existence of large resonance peak values, arithmetic averages are not adequate. A good choice is to select Σ_{ave} such that

$$\Sigma_{\text{ave}} \int_a^b \Phi(u) du = \int_a^b \Sigma(u)\Phi(u) du$$

or

$$f(\Sigma_{\text{ave}}) = \int_a^b f(\Sigma(u)) du$$

where f is a suitable function, e.g.,

$$f(\Sigma) = \frac{\Sigma}{\Sigma + \Sigma_d}$$

where Σ_d is carefully selected. (see [7]) ■

Roughly speaking, to compute *the resonance escape probability by using*(7.4), we should select u_0 in the fast neutrons domain ($u_0 < 7$) and u_1 in the thermal neutron domain ($u_1 > 16$).

We assume the reactor's core to be homogeneous, containing N_0 Uranium atoms per cm^3, and N_H hydrogen atoms per cm^3.

We note that for hydrogen, we have $\sigma_{\mathrm{sH}} = 20b$ in the slowing down lethargy interval (7, 16).

We shall neglect

- The hydrogen capture cross section (smaller in the slowing down domain than in the thermal neutron domain, the latter already appearing in the thermal utilization factor f).
- The Uranium elastic collision cross section (collisions with heavy nuclei like Uranium make the neutron change direction but energy changes are negligible).

Then we have

$$\Sigma_a(u) = N_0\sigma_U(u)$$
$$\Sigma_s(u) = N_H\sigma_{\mathrm{sH}}$$

and

$$\frac{\Sigma_a(u)}{\Sigma(u)} = \frac{N_0\sigma_U(u)}{N_0\sigma_U(u) + N_H\sigma_{\mathrm{sH}}} = \frac{\sigma_U(u)}{\sigma_U(u) + \sigma_d}$$

where we have introduce the *dilution ratio* N_H/N_0 and the *dilution cross section* defined by:

$$\sigma_d = \frac{N_H}{N_0}\sigma_{sH}.$$

The resonance escape probability p can be written as :

$$p = \exp\left(-\int_{u_0}^{u_1} \frac{\sigma_U(u')}{\sigma_U(u') + \sigma_d}\mathrm{d}u'\right) = \exp\left(-\frac{1}{\sigma_d}\int_{u_0}^{u_1} \sigma_U(u')\varphi(u')\mathrm{d}u'\right)$$

where we let

$$\varphi(u) = \frac{\sigma_d}{\sigma_U(u) + \sigma_d}$$

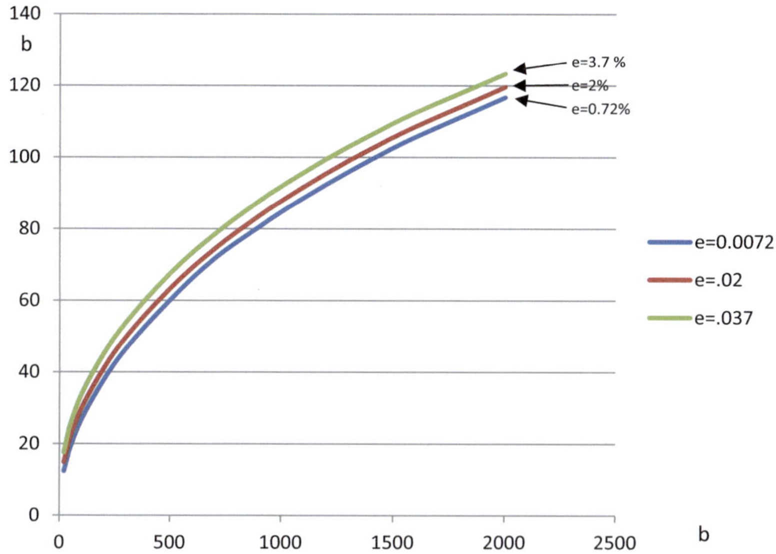

Fig. 7.2 Effective integral for uranium (3 enrichment values)

Finally, defining the effective integral (that depends on parameter σ_d)

$$I_{\text{eff}} = \int_{u_0}^{u_1} \sigma_U(u')\varphi(u')\mathrm{d}u'$$

We obtain

$$p = \exp\left(-\frac{1}{\sigma_d} I_{\text{eff}}\right) \tag{7.10}$$

which is exact for hydrogen.

For other moderators, the following formula can be used (see [7]) with $\sigma_d = \frac{N_m}{N_0}\sigma_{\text{sm}}$

$$p \cong \exp\left(-\frac{1}{\xi\sigma_d} I_{\text{eff}}\right) \tag{7.11}$$

The smaller ξ, the smaller p.

We note that I_{eff} is given in barns; it depends on σ_d that is also given in barns.

On Fig. 7.2 we plot the curve $\sigma_d \to I_{\text{eff}}$ for 3 different values of the enrichment e of uranium.

Table 7.1 Values of the slowing down parameters for various nuclei

Nucleus	A	γ	ε	ξ	n
H	1	0	∞	1	15
D	2	0.111	2.197	0.725	20
C	12	0.716	0.334	0.158	92
U	236	0.9832	0.017	0.0084	1717

Note that parameter n is the number of elastic collisions needed for slowing down

Note these are quite regular curves while the curve $u \to \sigma_U(u)$ is quite irregular (see Fig. 4.2).

On Table 7.1, we give values of the slowing down parameters for various nuclei; n is the number of elastic collisions needed for a fast neutron to become a thermal neutron.

The main moderators are light water (H), heavy water (D) and graphite (C): what is the best?

The following parameters should be taken into account (see Tab 7.1).

1. its atomic mass A, which leads to a greater or lesser value of ξ. On this point, hydrogen has a clear advantage.
2. its collision cross-section σ_{sm}: the larger it is, the more collisions there are. Here too, hydrogen has the advantage. (NB $\sigma_{\text{sH}} = 20b$ whereas $\sigma_{\text{sD}} = 3b$ and $\sigma_{\text{sG}} = 5b$).
3. its capture cross-section, which must be as small as possible. Here, however, graphite or heavy water have a clear advantage.

For countries having only access to natural uranium it is not possible to select light water as a moderator since its capture cross section is not small enough. France, e.g., has developed NUGG (natural uranium graphite gas cooled reactors) in the sixties. In the same way, India has built CANDU reactors having heavy water as a moderator.

Using enriched uranium rather than natural uranium makes it possible to use light water as a moderator and then to decrease the size of the core accordingly.

Exercise 7.3

1. Consider a fuel material made of 3.7% enriched uranium oxide (UO_2), density 11 g/cm^3.

   We denote by N_U the number of uranium atoms per cm^3, N_8 (resp. N_5) the number of ^{238}U (resp. ^{235}U) atoms per cm^3. Compute N_U, N_8 and N_5.
2. We select the following values for thermal neutron absorption cross sections $\sigma_8 = 2.7$ b for ^{238}U and $\sigma_5 = 680b$ for ^{235}U.

   We neglect the absorption for oxygen.

   Compute the absorption cross section Σ_{afuel} of the fuel
3. We specify $\sigma_{\text{aH}} = 0.322\ b$ for hydrogen absorption. Given the volumic mass 1g/cm^3 for light water used as a moderator, compute the number N_H of

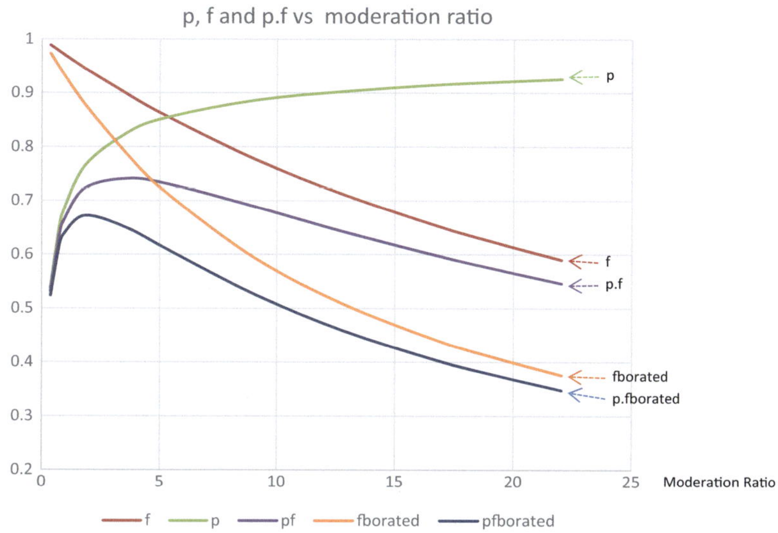

Fig. 7.3 Values of p, f, and $p.f$ w.r.t V_m/V_U (pure water vs borated water)

hydrogen atoms per cm^3 and the macroscopic absorption cross section Σ_{am} of the moderator.

4. We assume the core to be homogeneous: each cm^3 of the core contains ω cm^3 of UO_2 and $(1 - \omega)$ cm^3 of water. Compute the thermal utilization factor f when $\omega = 1/3$. ■ (*solution* p. 167)

When the moderation ratio $V_m/V_U = (1 - \omega)/\omega$ increases, then the thermal utilization factor f decreases, as can be seen on Fig. 7.3. On the contrary the resonance escape probability p increases.

Then, there is an optimal value for the moderation ratio, which is about equal to 4 when pure water is used as a moderator. However this is not the moderation ratio selected for the PWR reactors that is equal to 2.

The reason is connected to safety: a reactor with a liquid moderator should never be over-moderated. Indeed, a power increase would induce a temperature increase, and then a decrease of the moderator density and then, if the reactor is over-moderated a reactivity increase.

As can be seen on Fig. 7.3 the optimal value for the moderation ratio is not the same for pure water or for borated water. With 1000 ppm of Boron, the optimal value is more like $V_m/V_U = 2$.

This is the reason why the PWR designer selected a moderation ratio equal to 2.

Note: The RBMK that exploded in Chernobyl is an example of over-moderated reactor.

The graphite in the RBMK was sufficient to moderate neutrons. The steam generated by the boiling water was used to drive turbines. If the power increased, then

Table 7.2 Values of the resonance escape probability for 3 values of T

T (K)	r	$p = r^2$
293.16	0.943	0.89
1000	0.932	0.87
1800	0.927	0.86

boiling would increase, then the water density and the neutron capture by hydrogen would decrease, factor f would increase but p would not change. ■

Exercise 7.4 Case where the moderator is graphite containing only ${}^{12}_{6}C$ atoms. Knowing that the density of graphite is $1.6\,\mathrm{g/cm^3}$ compute the number N_G of ${}^{12}_{6}C$ atoms per cm^3. Given that $\sigma_{cG} = 0.0034\,b$ for thermal neutrons, compute the macroscopic cross section Σ_{cG} of graphite in the thermal domain.■ (solution p. 168)

Graphite therefore captures 100 times less thermal neutrons than light water. But is it a good moderator? Indeed, the mass $A = 12$ of the carbon atom means that neutrons are slowed down much less on each collision. A much higher moderation ratio is therefore required.

Enrico Fermi even realized that with natural Uranium, moderation had to be heterogeneous (not homogeneous) to obtain a multiplication coefficient $k_\infty > 1$.

Explanation of the Doppler Effect

Nuclear theory predicts that the energy dependence of the cross sections in the vicinity of each resonance will take the form of the Breit-Wigner formula. (see [2]).

$$\psi(E) = \frac{1}{2\beta\sqrt{\pi}} \int_{-\infty}^{+\infty} \frac{\exp(-(x-y)^2/4\beta^2)}{1+y^2} \mathrm{d}y \tag{7.12}$$

where

$x = 2\frac{E-E_0}{\Gamma}$ $\beta = \Delta/\Gamma$ $\Delta = \sqrt{4E_0 k_B T/A}$ k_B is Boltzmann's constant.

For the first and main resonance of ${}^{238}\mathrm{U}$, we have $E_0 = 6.674\,\mathrm{eV}$ and $\Gamma = 0.05$ eV.

We multiply ψ by a scaling factor and we plot $E \to \psi(E)$ for 3 values of T.

Then we compute the resonance escape probability induced by one resonance

$$r = \exp\left(-\int_{u_0}^{u_1} \frac{\psi(u')}{\psi(u') + \sigma_d} \mathrm{d}u'\right)$$

We give the results in Table 7.2 both for r (for one resonance), and r^2 (for two resonances)

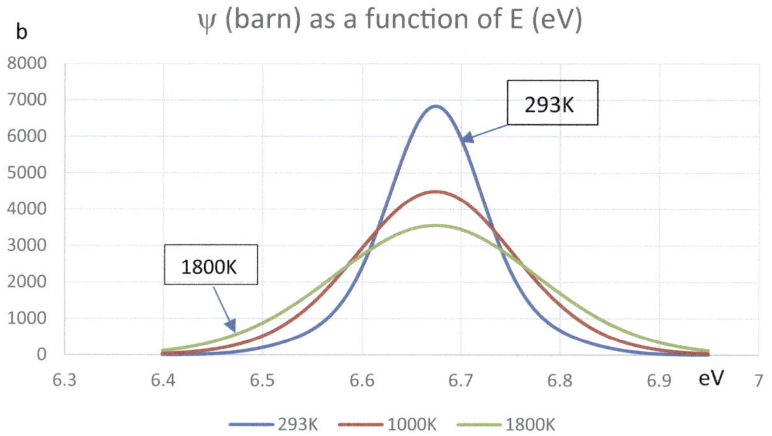

Fig. 7.4 Doppler broadening of the first ^{238}U resonance

Of course there are several clearly separated resonance peaks and for each of them the Doppler broadening is similar.

Assume there are only two of them, one from u_0 to u_1, and the second one from u_2 to u_3, we shall obtain

$$p = \exp\left(-\int_{u_0}^{u_3} \frac{\psi(u')}{\psi(u') + \sigma_d} \mathrm{d}u'\right) = \exp\left(-\int_{u_0}^{u_1} \frac{\psi(u')}{\psi(u') + \sigma_d} \mathrm{d}u'\right)$$

$$\cdot \exp\left(-\int_{u_2}^{u_3} \frac{\psi(u')}{\psi(u') + \sigma_d} \mathrm{d}u'\right) = r^2$$

Hence the 3rd column of Table 7.2. The broadening effect is illustrated on Fig. 7.4.

With such values of $p(T)$ the *reactivity* $\rho = (k - 1)/k$ decrease between 293 and 1000 K will be 2325 pcm, which gives about − 3 pcm/K. This is a very significant effect and we shall see in Chap. 15 that it helps stabilizing the reactor. (1 pcm $= 10^{-5}$)

Chapter 8
The Neutron Transport Equation and Its Diffusion Approximation

In this chapter, we introduce the linear Boltzmann equation that is solved in most neutron transport codes. We also introduce its diffusion approximation which is simpler to solve, and for which we shall give a few illustrative exercises. We follow references [1, 3].

After having studied the energy dependence of the neutron flux, we are going to study its spatial dependence (in x), but also in direction ($\overrightarrow{\Omega}$), because neutrons can go preferentially in certain directions.

The S_2 sphere and its parameterization.

In the following we shall work with unit vectors of $\boldsymbol{R}^3$, then vectors $\overrightarrow{\Omega}$ satisfying $\left|\overrightarrow{\Omega}\right| = 1$.

It is well known that in a given axis system we can represent $\overrightarrow{\Omega}$ by a polar angle $\theta \in [0, \pi]$ and by an azimuth angle $\varphi \in [0, 2\pi]$

When Oz is the polar axis, we have

$$\overrightarrow{\Omega} = \{\cos\varphi \cdot \sin\theta, \sin\varphi \cdot \sin\theta, \cos\theta\}$$

It is well known that, with such a parameterization, the surface element is $\mathrm{d}\Omega = \sin\theta \cdot \mathrm{d}\theta \cdot \mathrm{d}\varphi$ so that the integral of a function $f : \boldsymbol{S_2} \to \boldsymbol{R}$ satisfies:

$$\int\limits_{S_2} f(\Omega)\mathrm{d}\Omega = \int\limits_0^{2\pi} \mathrm{d}\varphi \int\limits_0^{\pi} f(\varphi, \theta) \sin\theta \, \mathrm{d}\theta$$

Incidentally, if we take $f = 1$, we find that the surface of the sphere is equal to 4π.

B. Mercier, *Simple Models for Understanding Nuclear Reactor Physics*,
https://doi.org/10.1007/978-3-031-98532-4_8

The notations used to establish the Boltzmann equation.

For the sake of simplicity, we place ourselves in the *monokinetic case*: let $N = N\left(x, \overrightarrow{\Omega}, t\right)$ denote the neutrons density at point $x \in \boldsymbol{R}^3$ traveling in direction $\overrightarrow{\Omega}$ at time t.

N is then measured in neutrons per cm^3 per steradian.

For historical reasons, neutron physicists do not work with neutron density N but with a quantity they call "flux":

$$\Phi\left(x, \overrightarrow{\Omega}, t\right) = vN\left(x, \overrightarrow{\Omega}, t\right)$$

where v is the speed of neutrons.

Φ is then measured in neutrons per cm^2 per steradian per second.

Let $V \subset \boldsymbol{R}^3$ denote an arbitrary bounded volume with boundary Γ assumed to be regular, so that it is possible to define a surface element dS.

We denote by $\overrightarrow{n} = \overrightarrow{n}(x)$ the unit normal vector in $x \in \Gamma$ outwards directed w.r.t to V.

Definitions and properties of the divergence operator

If we denote by $\{q_1, q_2, q_3\}$ the components of vector $\vec{q}$, the divergence operator is defined by:

$$\text{div } \vec{q} = \frac{\partial q_1}{\partial x_1} + \frac{\partial q_2}{\partial x_2} + \frac{\partial q_3}{\partial x_3} \tag{8.1}$$

it is therefore a differential operator involving partial derivatives.

In Wikipedia, http://en.wikipedia.org/wiki/Divergence_theorem we find (with other notations) the following statement:

Divergence theorem (also called Ostrogradsky's theorem)

Let V denote a domain in $\boldsymbol{R}^3$ with boundary Γ, and $\vec{q} = \{q_1, q_2, q_3\}$ then we have (provided that the functions appearing in the integrals are integrable): (where $\overrightarrow{n}$ denotes the unit normal vector to V outwards directed)

$$\int_V \text{div } \overrightarrow{q} \, \text{d}x = \int_\Gamma \overrightarrow{q} \cdot \overrightarrow{n} \, \text{d}S \tag{8.2}$$

■

The unsteady Boltzmann equation

We will use this result to establish the Boltzmann equation.

We give ourselves a volumetric source of neutrons $S = S\left(x, \overrightarrow{\Omega}, t\right)$ measured in neutrons/steradian/s. We assume an absorbing medium and we denote by $\Sigma = \Sigma(x)$

the macroscopic absorption cross section, assumed isotropic, i-e not depending on $\vec{\Omega}$.

Let us make a neutron balance in our arbitrary volume V.

At time t the number of neutrons traveling in a direction located in a small solid angle $\mathrm{d}\Omega$ around $\vec{\Omega}$ in $x \in \boldsymbol{R}^3$ is equal to

$$\mathrm{d}\Omega \int_V N\left(x, \vec{\Omega}, t\right)\mathrm{d}x = \frac{1}{v}\mathrm{d}\Omega \int_V \Phi\left(x, \vec{\Omega}, t\right)\mathrm{d}x.$$

(With our notations, the number of neutrons located in V at time t is equal to

$$\frac{1}{v}\int_{S_2} \mathrm{d}\vec{\Omega} \int_V \Phi\left(x, \vec{\Omega}, t\right)\mathrm{d}x)$$

The number of neutrons traveling at time t in direction $\vec{\Omega}$ (more precisely having a direction located in a small solid angle $\mathrm{d}\Omega$ around $\vec{\Omega}$) can increase or decrease:

- by the fact that source S is producing neutrons
- by the fact that the medium is absorbing neutrons.

In Chap. 2, we have seen that the number of neutrons traveling in a direction located in the solid angle $\mathrm{d}\Omega$ around $\vec{\Omega}$ that are absorbed each second in V is:

$$\mathrm{d}\Omega \int_V \Sigma\Phi\left(x, \vec{\Omega}, t\right)\mathrm{d}x$$

- by the fact that there are both incoming and outgoing neutrons in domain V

Let $\Gamma_+ = \Gamma_+(\Omega) = \left\{x \in \Gamma : \vec{\Omega}.\vec{n} \geq 0\right\}$ and $\Gamma_- = \Gamma_-(\Omega) = \left\{x \in \Gamma : \vec{\Omega} \quad \vec{n} < 0\right\}$ we let

$J_+ = \int_{\Gamma_+} \vec{\Omega} \cdot \vec{n}\, \Phi\, \mathrm{d}S$ and $J_- = -\int_{\Gamma_-} \vec{\Omega} \cdot \vec{n}\, \Phi\, \mathrm{d}S$

So that J_+ can be interpreted as the *outgoing neutronic current* from V and J_- as the *incoming neutronic current* in V.

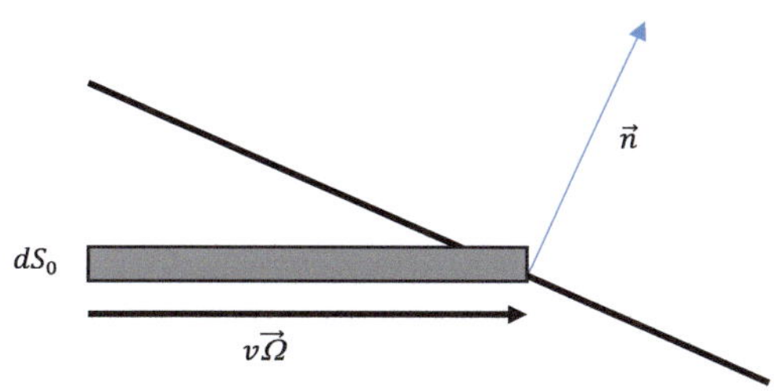

There are $Nv\mathrm{d}S_0$ neutrons in the gray rectangle. All these neutrons travel in direction $\vec{\Omega}$.

In one second these neutrons cross the surface element $\mathrm{d}S$ located on the black line.

We obviously have $\mathrm{d}S_0 = \mathrm{d}S \cos\theta$ where $\cos\theta = \vec{\Omega} \cdot \vec{n}$ which establishes our interpretation.

Moreover: $J = \int_\Gamma \vec{\Omega} \cdot \vec{n}\, \Phi \,\mathrm{d}S = J_+ - J_-$ is called the *net outgoing current* from V.

Globally, we have then

$$\frac{1}{v}\frac{d}{\mathrm{d}t}\int_V \Phi\left(x, \vec{\Omega}, t\right)\mathrm{d}x + \int_\Gamma \vec{\Omega}.\vec{n}\, \Phi \,\mathrm{d}S + \int_V \Sigma\Phi\left(x, \vec{\Omega}, t\right)\mathrm{d}x = \int_V S\left(x, \vec{\Omega}, t\right)\mathrm{d}x$$

Applying the divergence theorem, we obtain:

$$\int_V \left[\frac{1}{v}\frac{\partial}{\partial t}\Phi\left(x, \vec{\Omega}, t\right) + \mathrm{div}\left(\vec{\Omega}\,\Phi\right) + \Sigma\Phi\left(x, \vec{\Omega}, t\right)\right]\mathrm{d}x = \int_V S\left(x, \vec{\Omega}, t\right)\mathrm{d}x$$

and, since V is arbitrary, we deduce the Boltzmann equation:

$$\frac{1}{v}\frac{\partial}{\partial t}\Phi + \mathrm{div}\left(\vec{\Omega}\,\Phi\right) + \Sigma\Phi = S;$$

It is true in all point x, all direction $\vec{\Omega}$ at each time t.

By using the repeated index summation convention we can also write

$$\frac{1}{v}\frac{\partial}{\partial t}\Phi + \Omega_i\frac{\partial}{\partial x_i}\Phi + \Sigma\Phi = S \tag{8.3}$$

Steady case

When the source S is independent of time, if we start from $\Phi = 0$ at time $t = 0$, or from any other value of the neutron flux Φ, it is conceivable that, after a transitional period, and in practice by making $t \to \infty$ in (8.3), we eventually stabilize toward a stationary state, still noted Φ but which no longer depends on time.

In other words $\frac{\partial\Phi}{\partial t} = 0$ and Φ satisfies the steady Boltzmann equation:

$$\Omega_i\frac{\partial}{\partial x_i}\Phi + \Sigma\Phi = S \quad x \in V,\ \Omega \in \boldsymbol{S_2} \tag{8.4}$$

Isotropic collisions in the laboratory reference frame.

In principle, as we have seen in Annex 1, the collision term is isotropic in the Center of Mass reference frame. To obtain more easily the diffusion approximation of the Boltzman equation, we shall make the simplifying assumption that the collision term is isotropic in the laboratory reference frame. The cross section Σ that appears in (8.4) then satisfies $\Sigma = \Sigma_0 + \Sigma_1$ where Σ_0 is the absorption cross section and Σ_1 the elastic scattering cross section.

The events undergone by neutrons moving in the direction $\vec{\Omega}$ may correspond (with probability Σ_0/Σ) to absorption or (with probability Σ_1/Σ) to an elastic collision which causes the neutrons to reappear in another direction $\vec{\Omega}'$ uniformly distributed on the sphere S_2.

The source S in Eq. (5.4) is now of the form

$$S = S_0 + \frac{1}{4\pi}\Sigma_1 \int_{S_2} \Phi\left(x, \vec{\Omega}'\right) d\vec{\Omega}'$$

where $S_0 = S_0(x)$ is the true neutron source (assumed isotropic for simplicity) and where the second term takes into account the fact that neutrons previously going in direction $\vec{\Omega}'$ reappear (with a probability density equal to $1/4\pi$) in direction $\vec{\Omega}$.

We can then write:

$$\Omega_i \frac{\partial}{\partial x_i}\Phi + \Sigma\Phi = S_0 + \frac{1}{4\pi}\Sigma_1 \int_{S_2} \Phi\left(x, \vec{\Omega}'\right) d\vec{\Omega}' \tag{8.5}$$

Integrating for $x \in V$ and $\Omega \in \boldsymbol{S_2}$, we deduce:

$$\int_{S_2} d\Omega \int_{\Gamma} \vec{\Omega} \cdot \vec{n}\, \Phi\, dS + \int_{S_2} d\Omega \int_{V} \Sigma \cdot \Phi dx$$
$$= 4\pi \int_{V} S_0 dx + \int_{V} \Sigma_1 dx \int_{S_2} \Phi\left(x, \vec{\Omega}'\right) d\vec{\Omega}'$$

i-e:

$$\int_{S_2} d\Omega \int_{\Gamma} \vec{\Omega} \cdot \vec{n}\Phi dS + \int_{S_2} d\Omega \int_{V} \Sigma_0 \cdot \Phi\, dx = 4\pi \int_{V} S_0 dx \tag{8.6}$$

which shows that the source is equal to absorption plus net current, i.e., if there are no neutrons coming in: source = absorption + leakage.

Boundary conditions for the Boltzmann equation

The Boltzmann equation (8.4) alone is not enough to define a unique solution Φ. Boundary conditions must also be defined, for example, giving the incoming flux.

More precisely we can prescribe

$$\Phi(x,\Omega) = \Phi_d(x,\Omega) \quad \text{if } x \in \Gamma \text{ and } \Omega \cdot n(x) < 0 \tag{8.7}$$

where Φ_d is given.

We can alternatively prescribe some reflexive conditions like:

$$\Phi(x,\Omega) = \alpha\Phi\left(x,\Omega'\right) \quad \text{if } x \in \Gamma \text{ and } \Omega \cdot n(x) < 0 \text{ with } \Omega' = \Omega - 2(\Omega \cdot n)n$$

where α is an albedo coefficient such that $0 < \alpha \leq 1$ ($1 - \alpha$ is an attenuation coefficient).

It can be proved (see e.g., Dautray-Lions [8]) that system (8.5) (8.7) is a well posed problem, which means that it has a unique solution if $\Sigma_0 > 0$ and $\Sigma_1 \geq 0$.

Note that the solution Φ to problem (8.5) and (8.7) is a function of 5 variables (3 spatial variables x, y, z and 2 angular variables $\theta,\ \varphi$).

Such a problem can be solved by the Monte-Carlo method. A deterministic method is more difficult. That's why people often use the diffusion approximation of the Boltzmann equation, that we shall describe later on.

Special case: plane symmetry.

In the above, Σ_0 and Σ_1 may depend on $\boldsymbol{x} = \{x, y, z\}$. If we assume that:

- the domain V is located between the planes $z = a$ and $z = b$, with $a < b$
- $\Sigma_0 = \Sigma_0(z);\ \Sigma_1 = \Sigma_1(z);\ S_0 = S_0(z)$
- The boundary data $\Phi_d(x,\Omega)$ is such that Φ_d depends only on μ when $\boldsymbol{x} = \{x, y, a\}$ or $\boldsymbol{x} = \{x, y, b\}$ where $\mu = \cos\theta$ and θ is the polar angle, i-e the angle between direction Ω and the Oz axis,

Then, a symmetry argument shows that $\Phi = \Phi(z,\mu)$ i-e flux Φ is depending on z and μ only.

Moreover, let $\Sigma = \Sigma_0 + \Sigma_1$ equation (8.5) writes:

$$\begin{aligned}
\mu\frac{\partial\Phi}{\partial z} + \Sigma\Phi &= S_0 + \frac{1}{4\pi}\Sigma_1 \int\limits_{S_2} \Phi\left(x, \overrightarrow{\Omega'}\right) \mathrm{d}\overrightarrow{\Omega'} \\
&= S_0 + \frac{1}{4\pi}\Sigma_1 \int\limits_0^{2\pi} \mathrm{d}\varphi' \int\limits_{-1}^{1} \Phi\left(z,\mu'\right)\mathrm{d}\mu'
\end{aligned}$$

i-e

$$\mu \frac{\partial \Phi}{\partial z} + \Sigma \Phi = S_0 + \frac{1}{2}\Sigma_1 \int_{-1}^{1} \Phi(z, \mu')\mathrm{d}\mu' \quad \text{for } a \le z \le b \text{ and } -1 \le \mu \le 1 \quad (8.8)$$

We may complement (8.8) with the boundary conditions:

$$\Phi(a, \mu) = 0 \ \text{ for } \mu > 0 \ \text{ and } \Phi(b, \mu) = A \text{ for } \mu < 0. \quad (8.9)$$

i-e the incoming flux is given.

System (8.8) and (8.9) is also a well posed problem, if $\Sigma_0 > 0$ and $\Sigma_1 \geq 0$.

Its solution Φ is a function of 2 variables only: the space variable z and the angular variable μ. ■

Diffusion approximation of the monokinetic Boltzmann equation

For the diffusion approximation to be valid, the mean free path $1/\Sigma$ must be small, and the medium should be homogeneous, but that's not enough: the effective scattering cross-section must dominate over the absorption cross-section.

i-e $\Sigma_s \gg \Sigma_0$.

For light water, we have $\Sigma_s = 1.34\,\mathrm{cm}^{-1}$ and $\Sigma_0 = 0.02\,\mathrm{cm}^{-1}$, so the diffusion approximation is valid if the size of the domain is greater than 10 cm. In the core of a PWR where the moderation ratio is about equal to 2, it is not so true since for thermal neutrons we typically have $\Sigma_0 \sim 0.5\,\mathrm{cm}^{-1}$ in the fuel, but the diffusion approximation is often used since it gives acceptable results.

In Annex 2, we show how to obtain the diffusion equation:

$$-\mathrm{div}(D.\nabla\Phi_0) + \Sigma_0\Phi_0 = S_0 \quad \text{pour } x \in V \quad (8.10)$$

and the Fourier boundary condition

$$\frac{1}{4}\Phi_0 + \frac{D}{2}\frac{\partial}{\partial n}\Phi_0 = 0 \quad (8.11)$$

on the boundary Γ.

Coefficient D that appears in (8.10) and (8.11) is called diffusion coefficient.

In Annex 2, with simplifying hypotheses, it is shown that

$$D = \frac{1}{3.\Sigma} \quad (8.12)$$

So that the diffusion coefficient is small if the mean free path is small.

Note that we may have $D = D(x)$ when $\Sigma = \Sigma(x)$.

It is well known that system (8.10) and (8.11) has a unique solution if $D(x) \geq D_0 > 0$ and $\Sigma_0 \geq 0$.

We can see that solution Φ_0 of (8.10) and (8.11) depends on x only, then on the 3 space variables x, y, z. Thus, it is quite easy to compute Φ_0 with deterministic methods such as the finite element method.

However, it must be borne in mind that if we introduce the current

$$\vec{J} = -D \cdot \nabla \Phi_0 \tag{8.13}$$

It is $\Phi_0(x) + 3\vec{\Omega} \cdot \vec{J}(x)$ that is a good approximation to $\Phi(x, \Omega)$; not $\Phi_0(x)$.

Relation (8.13) is well known as "Fick's law".

Note that, mathematically, there is a good analogy between neutronics (Fick's law) and thermics (Fourier's law), but current $\vec{J}$ (not the neutron flux) is the equivalent of the thermal flux.

Remark 8.1 Taking anisotropy into account.

If we were to take into account the anisotropy of collisions in the laboratory reference frame (without calling into question the isotropy in the center of mass), we would obtain (see [2]):

$$D = \frac{1}{3 \cdot \Sigma_{tr}}$$

where $\Sigma_{tr} = \Sigma - \overline{\mu}\,\Sigma_s$ and $\overline{\mu}$ is the mean value of the cosine of the deviation angle ($\overline{\mu} = 2/3$ for hydrogen).

In the case where the atomic mass of the moderator nuclei is $A > 1$ the formula $\overline{\mu} \cong \frac{2}{3A}$ can be used (see [2]). ■

Interpretation of the Fourier boundary condition (8.11).

If, as it should, D is small with respect to the size of domain V, for $x \in \Gamma$, the point $x + 2D\vec{n}$ is not far from the boundary Γ. Let

$$f(s) = \Phi_0\big(x + s\vec{n}\big),$$

We have $f(s) \cong f(0) + s.f'(0)$

Since $f'(0) = \frac{\partial}{\partial n}\Phi_0$, the equation $\Phi_0 + 2D\frac{\partial}{\partial n}\Phi_0 = 0$ is equivalent to $f(2D) \cong 0$ which is why we call 2D the *extrapolation distance*. See Fig 8.1.

Everything happens as if Φ_0 were not zero at the boundary, but at a distance 2D from the boundary.

Taking Φ_0 to be zero at the boundary, as we will do in Chap. 9, means committing a small error, of which we know the order of magnitude. ■

Exercise 8.1 Let $\Phi = \Phi(z, \mu)$ denote the solution of the 1D Boltzmann equation

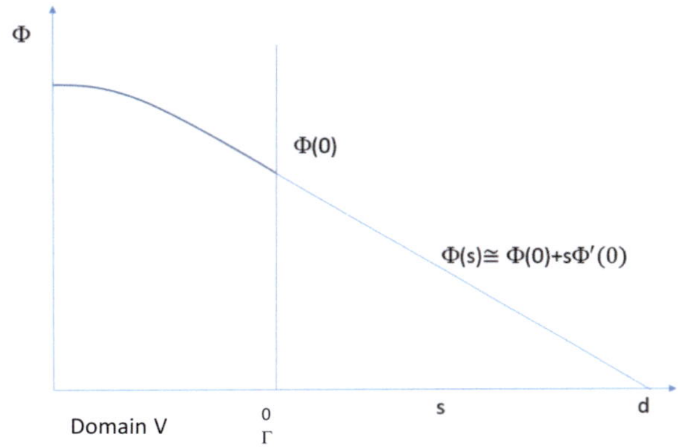

Fig. 8.1 Extrapolation distance $d = 2D$

$$\mu \frac{\partial \Phi}{\partial z} + \Sigma \Phi = S_0 + \frac{1}{2}\Sigma_s \int_{-1}^{1} \Phi(z, \mu') \mathrm{d}\mu' \quad \text{for}\, 0 \le z \le L \,\text{and} \, -1 \le \mu \le 1 \tag{8.14}$$

satisfying the following boundary condition:

$$\Phi(0, \mu) = A \,\text{for}\, \mu > 0 \,\text{and}\, \Phi(L, \mu) = 0 \,\text{for}\, \mu < 0. \tag{8.15}$$

We take for true that this problem has a unique solution if $\Sigma = \Sigma_s + \Sigma_0$ with $\Sigma_0 > 0$ et $\Sigma_s \ge 0$.

Like in the general case, we shall replace $\Phi(z, \mu)$ by $\Phi_0(z) + 3\mu J(z)$ in (8.14). As $\Phi_0(z) + 3\mu J(z)$ is not rigorously equal to Φ, we shall obtain a residue

$$F = \mu \frac{\partial}{\partial z}(\Phi_0 + 3\mu J) + \Sigma(\Phi_0 + 3\mu J) - S_0 - \frac{1}{2}\Sigma_s \int_{-1}^{1} (\Phi_0 + 3\mu' J) \mathrm{d}\mu'$$

that will not vanish rigorously.

We require that for all $z \in [0, L]$

$$\int_{-1}^{1} F \mathrm{d}\mu = 0$$

$$\int_{-1}^{1} F \mu \mathrm{d}\mu = 0$$

1. Prove that

$$F = \mu \frac{\partial}{\partial z}(\Phi_0 + 3\mu J) + \Sigma(\Phi_0 + 3\mu J) - S_0 - \Sigma_s \Phi_0$$

2. Prove that

$$\frac{d}{dz} J + \Sigma_0 \Phi_0 = S_0. \tag{8.16}$$

3. Prove that:

$$\frac{d}{dz} \Phi_0 + 3\Sigma J = 0 \tag{8.17}$$

4. We let

$$J_+(z) = \frac{1}{2} \int_0^1 \Phi(z, \mu)\mu \, d\mu$$

$$J_-(z) = -\frac{1}{2} \int_{-1}^0 \Phi(z, \mu)\mu \, d\mu.$$

Prove that

$$J_+ = \frac{1}{4}\Phi_0(z) + \frac{1}{2}J(z) \tag{8.18}$$

$$J_- = \frac{1}{4}\Phi_0(z) - \frac{1}{2}J(z) \tag{8.19}$$

5. Prove that Φ_0 satisfies the diffusion equation

$$-D \frac{d^2}{dz^2} \Phi_0 + \Sigma_0 \Phi_0 = S_0 \text{ for } 0 \leq z \leq L$$

with $D = \frac{1}{3\Sigma}$ and

$$\frac{1}{4}\Phi_0(0) - \frac{D}{2}\left(\frac{d}{dz}\Phi_0\right)(0) = \frac{1}{4}A$$

$\frac{1}{4}\Phi_0(L) + \frac{D}{2}\left(\frac{d}{dz}\Phi_0\right)(L) = 0.$ ■ (*solution* p. 168)

Exercise 8.2 A slab of light water $0 < z < L$ infinite in x and y directions is given.

It is assumed that the problem can be treated in plane symmetry, and that the diffusion approximation (8.16) and (8.17) is valid.

Here, we denote by Φ what we called Φ_o in the previous exercise, so that we have

$$\frac{\mathrm{d}J}{\mathrm{d}z} + \Sigma_a \Phi(z) = 0$$
$$J(z) = -D\Phi'(z) \tag{8.20}$$

We also let

$$J_+ = \frac{1}{4}\Phi(z) + \frac{1}{2}J(z)$$
$$J_- = \frac{1}{4}\Phi(z) - \frac{1}{2}J(z)$$

So that the net current J is the difference between J_+ (left to right current) and J_- (right to left current): $J = J_+ - J_-$.

The incoming current $J_+(0) = A$, is supposed to be given. At the other end, no neutrons are incoming so that $J_-(L) = 0$. The diffusion coefficient D and the macroscopic absorption cross section Σ_a for light water are also assumed to be given.

1. Let $\beta = \sqrt{\Sigma_a/D}$, prove that

$$\Phi(z) = a\, ch(\beta z) + b\, sh(\beta z)$$

2. Compute $J_+(z)$ and $J_-(z)$. Deduce 2 equations satisfied by parameters a and b.
3. Numerical application: let $L = 50$ cm, $D = 0.27$ cm and $\sum_a = 0.02\,\text{cm}^{-1}$.
   $A = 50\,\text{n/cm}^2/\text{s}$, compute $J_+(L)$ and $J_-(0)$
4. Calculate the absorption in a volume of $L\,\text{cm} \times 1\,\text{cm}^2$ of this supposedly infinite slice of water in the x and y directions.■ (*solution* p. 169)

Exercise 8.3 We now take the case where the reflector is made of heavy water, for which we can assume $\Sigma_a = 0$. As in the previous exercise, we give the incoming current $J_+(0) = A$ and $J_-(L) = 0$. The diffusion coefficient D (not the same as for light water) is assumed to be known.

1. Prove that

$$J_-(0) = \frac{1}{4}\frac{J}{D}L \quad \text{and} \quad J_+(0) = J + \frac{1}{4}\frac{J}{D}L = J\left(1 + \frac{L}{4D}\right)$$

   where J is a constant number. Find a relation between J, A, L and D.
2. We specify that $\sigma \cong 10b$ is the microscopic scattering cross section for heavy water and that $N_{D2O} = 0.0332 \times 10^{24}\,\text{molecules/cm}^3$. Compute the associated

macroscopic cross section Σ, the diffusion coefficient D and then $J_-(0)$ as a function of A, L and D. Take $A = 50\,\mathrm{n/cm^2/s}$ and evaluate $J_-(0)$. Is heavy water a good reflector ?■ (*solution* p. 171)

Chapter 9
One Group Diffusion Model

In the previous chapter, we have assumed that the neutron source is given. This is not the case in the core of a nuclear reactor, since the fission source depends on the neutron flux. Mathematically, we obtain an *eigenvalue problem that* we shall be able to solve in a few particular cases. From such an eigenvalue problem, we shall be able to deduce a *critical equation*, at least in the homogeneous case.

In a way, the one group diffusion model is the simplest mathematical model for the core of a nuclear reactor. We shall also introduce the *migration area*, and show its link with the non-leakage probability.

Starting from the monokinetic Boltzmann equation, we have derived the following problem (Note that we shall neglect the extrapolation distance.):

$$-\operatorname{div}(D \cdot \nabla_0 \Phi_0) + \Sigma_0 \Phi_0 = S_0 \quad \text{for}\, x \in V \tag{9.1}$$

$$\Phi_0 = 0 \text{ on the boundary } \Gamma. \tag{9.2}$$

We know that this problem has a unique solution $\Phi_0 \geq 0$. when $\Sigma_0 > 0$ (or alternatively, when $\Sigma_0 = \Sigma_0(x)$, and $\Sigma_0(x) \geq \Sigma_{\min} > 0$).

When we apply this model to PWRs, we are particularly interested in the group of thermal neutrons, whose energy is around 0.05 eV.

It's true that there are fast neutrons in the core of a PWR, but due to the presence of a moderator they are slowed down fairly quickly and then thermalized. Moreover, the fission cross section ^{235}U being much greater for thermal neutrons, most fissions occur in the thermal domain.

Remark 9.1 Table 9.1 is reproduced from [9, p. 382]: it corresponds to what we have in the fuel pins of a PWR 1300. For 4 groups of neutrons we give its upper and lower energy limit (in MeV). The 4 neutron fluxes are integral values from the lower to the upper bound.

Group 4 corresponds to thermal neutrons.

B. Mercier, *Simple Models for Understanding Nuclear Reactor Physics*,
https://doi.org/10.1007/978-3-031-98532-4_9

Table 9.1 Neutron flux (n/cm^2/s) and fission/cm^3/s in 4 groups of neutrons

	Gr 1	Gr 2	Gr 3	Gr 4	
Higher energy	10	0.90718	2.00E−03	6.25E−07	MeV
Lower energy	0.90718	2.00E−03	6.25E−07	1.1E−10	MeV
Neutron flux	7.51E+13	1.14E+14	6.43E+13	3.66E+13	n/cm^2/s
Fission/cm^3/s	8.29E+10	1.91E+11	1.58E+12	9.00E+12	

If we call "fast" the neutrons of the 3 other groups, then the "fast" neutron flux is 7 times higher than the thermal neutron flux. Despite this, 80% of fissions are produced by thermal neutrons. ■

In the diffusion equation above, Σ_0 is the macroscopic absorption cross section; it is the sum of two terms: $\Sigma_0 = \Sigma_c + \Sigma_f$.

Indeed absorption = capture + fission.

We can see that some of the thermal neutrons absorbed will give rise to fissions, producing a source term proportional to Φ_0. In the absence of external sources, we feel like writing $S_0 = \nu\Sigma_f\Phi_0$ but this is a source term in the fast range, and it should not be forgotten that, during the slowing down phase, a proportion $(1-p)$ of these fast neutrons will be captured by U^{238}, where p is the resonance escape probability. Then we should introduce the following source term $S_0 = p\cdot\nu\Sigma_f\Phi_0$ in our diffusion equation, or even better $S_0 = \varepsilon\cdot p\cdot\nu\Sigma_f\Phi_0$ to take fast fissions into account.

Note that Fermi's fast fission factor appears in the latter.

Finally, let $\nu_e = \varepsilon\cdot p\cdot\nu$, $S_0 = \nu_e\Sigma_f\Phi_0$. is the source term to be introduced in our diffusion on that now writes:

$$-\operatorname{div}(D\cdot\nabla\Phi_0) + \Sigma_0\Phi_0 = \nu_e\Sigma_f\Phi_0 \quad \text{for}\, x\in V. \tag{9.3}$$

$$\Phi_0 = 0 \quad \text{on } \Gamma. \tag{9.4}$$

We shall see that, actually, this is an eigenvalue problem.

Remark 9.2 (*Finite dimensional analog*) The finite dimensional analog to (9.1) and (9.2) would be to look for a vector $\Phi_0 \in \boldsymbol{R}^N$ such that

$$A\Phi_0 = S_0 \tag{9.5}$$

where A would be a $N\times N$ matrix.

For the existence and uniqueness of a solution to problem (9.5), we know that a necessary and sufficient condition is that the matrix A be invertible.

The finite dimensional analog to (9.3) and (9.4) would rather be:

$$A\Phi_0 = \mu_0\Phi_0 \tag{9.6}$$

where $\mu_0 \in \boldsymbol{R}$ is given.

We know that the solution to (9.6) is $\Phi_0 = 0$ *except* if μ_0 is an eigenvalue of matrix A, in which case Φ_0 is the eigenvector associated to μ_0; provided that μ_0 is a simple eigenvalue, Φ_0 is then defined up to a multiplicative factor. (In the kind of applications we have in mind all the eigenvalues will be positive and we are interested by the smallest one. Another nice property is that the matrix $A^{-1} \geq 0$ so that, from the Perron–Frobenius theorem (cf. Wikipedia), the eigenvector Φ_0 associated to the smallest eigenvalue μ_0 will be such that $\Phi_0 \geq 0$). ■

Coming back to (9.3) and (9.4) we can write

$$-\operatorname{div}(D \cdot \nabla\Phi_0) = \mu_0\Phi_0 \quad \text{for } x \in V \tag{9.7}$$

$$\Phi_0 = 0 \text{ on } \Gamma. \tag{9.8}$$

together with

$$\mu_0 = \nu_e\Sigma_f - \Sigma_0 \tag{9.9}$$

And we can even require $\Phi_0 \geq 0$ so that μ_0 will be the smallest eigenvalue from the Krein–Rutman theorem (see [8]).

Let's summarize: to solve the one group diffusion model (9.3) and (9.4) we have to

- solve the eigenvalue problem (9.7) and (9.8) and determine its smallest eigenvalue $\mu_0 > 0$
- check that the associated eigenfunction $\Phi_0 \geq 0$ on V
- adjust Σ_0 so that the critical equation (9.9) is satisfied (in a PWR, this can be achieved by tuning the Boron concentration in the light water).

In an equivalent way we can compute the effective multiplication factor $k = \frac{\nu_e\Sigma_f}{\mu_0+\Sigma_0}$ and require $k = 1$ for criticality.

Remark 9.3 *Interpretation of* (9.3) *and* (9.4) *as the limit as* $t \to \infty$. *of some unsteady problem.* Let us consider the following differential system:

$$\frac{\mathrm{d}}{\mathrm{d}t}\Phi + A\Phi = \mu\Phi. \tag{9.10}$$

where

- A is a symmetric $N \times N$ positive definite matrix
- $\Phi = \Phi(t) \in \boldsymbol{R}^N$
- $\mu > 0$.

To have a unique solution (9.10) should be complemented by an initial condition:

$$\Phi(0) = \psi$$

where $\psi \in \boldsymbol{R}^N$.

Let us introduce the eigenvalues $0 < \mu_0 < \mu_1 \leq \ldots \leq \mu_{N-1}$ and the eigenvectors $\{\Phi_i\}$, $0 \leq i \leq N-1$ of A defined by

$$A\Phi_i = \mu_i \Phi_i$$

Let us define $\xi_i(t)$, $i = 0, \ldots, N-1$ such that

$$\Phi(t) = \sum_{0}^{N-1} \xi_i(t)\Phi_i.$$

From (9.10) we deduce

$$\frac{\mathrm{d}}{\mathrm{d}t}\xi_i = (\mu - \mu_i)\xi_i$$

and

$$\xi_i(t) = \xi_i(0)\exp((\mu - \mu_i)t)$$

When we increase μ after starting from zero, we see that, while $0 \leq \mu < \mu_0$ all the coefficients $\xi_i(t)$ decrease exponentially with t. In the limit case $\mu = \mu_0$ all the $\xi_i(t)$ still decrease exponentially *but* $\xi_0(t)$ will remain constant, or even increase exponentially if $\mu > \mu_0$ (even if $\mu < \mu_1$).

This is why Φ_0 associated to the smallest eigenvalue μ_0 is called the *dominant mode*; μ_0 is called the *critical* eigenvalue; when $0 \leq \mu < \mu_0$ the system is said to be *sub-critical*, whereas, when $\mu > \mu_0$ it is *supercritical*. ■

Remark 9.4 (*Interpretation of the critical equation*(9.9)) In the same way, the solution $\Phi = \Phi(x, t)$ of the unsteady problem

$$\alpha\frac{\partial\Phi}{\partial t} - \mathrm{div}(D \cdot \nabla\Phi) + \Sigma_0\Phi = \nu_\mathrm{e}\Sigma_\mathrm{f}\Phi, \; x \in V, \; t > 0 \tag{9.11}$$

$$\Phi = 0, \; x \in \Gamma, \; t > 0 \tag{9.12}$$

should converge as $t \to \infty$ to the solution of (9.3) and (9.4).

However, this is true only in the sub-critical case $\frac{\nu_\mathrm{e}\Sigma_\mathrm{f}}{\mu_0+\Sigma_0} < 1$ or critical case $\frac{\nu_\mathrm{e}\Sigma_\mathrm{f}}{\mu_0+\Sigma_0} = 1$.

In the supercritical case $\frac{\nu_\mathrm{e}\Sigma_\mathrm{f}}{\mu_0+\Sigma_0} > 1$ it can be easily proved that $\Phi(x, t) \to \infty$ as $t \to \infty$. ■

Remark 9.5 (*Homogeneous case*) In this case Σ_0, Σ_f and D do not depend on the space variable x.

We let $k_\infty = \nu_e \Sigma_f / \Sigma_0$. Since $\nu_e = \varepsilon \cdot p \cdot \nu$, we have

$$k_\infty = \varepsilon \cdot p \cdot \frac{\nu \Sigma_f}{\Sigma_0} = \varepsilon \cdot p \cdot \frac{\nu \Sigma_f}{\Sigma_U} \frac{\Sigma_U}{\Sigma_0} = \varepsilon \cdot p \cdot f \cdot \eta$$

where $f = \frac{\Sigma_U}{\Sigma_0}$ is the thermal utilization factor.

Note that Σ_U, appearing in the previous formulae, stands for the macroscopic absorption cross section of Uranium in the core. Then, we have $\Sigma_0 = \Sigma_U + \Sigma_H$ where Σ_H is the macroscopic absorption cross section of hydrogen, that is not negligible in the thermal domain (for oxygen it is negligible).

Let $M^2 = \frac{D}{\Sigma_0}$, to compute the effective multiplication factor k we just have to solve:

$$-M^2 \Delta\Phi + \Phi = \frac{k_\infty}{k}\Phi \quad \text{for } x \in V \tag{9.13}$$

$$\Phi(x) \geq 0 \quad \text{for } x \in V \tag{9.14}$$

$$\Phi = 0 \quad \text{on the boundary } \Gamma \tag{9.15}$$

Or, equivalently, we can solve

$$-\Delta\Phi = \lambda_0 \Phi \quad \text{for } x \in V \tag{9.16}$$

$$\Phi(x) \geq 0 \quad \text{for } x \in V \tag{9.17}$$

$$\Phi = 0 \quad \text{on } \Gamma. \tag{9.18}$$

And then compute

$$k = \frac{k_\infty}{1 + \lambda_0 M^2} \tag{9.19}$$

We have seen that in the case where the domain V is an infinite slab in the x and y directions but bounded in z by $0 \leq z \leq H$ then $\lambda_0 = (\pi/H)^2$.

The parameter M^2 is called the migration area. Up to a factor 6 it can be interpreted see [2, 3] as the mean square of the crow-fly distance traveled by the neutron from emission to absorption.

It appears that $1/(1 + \lambda_0 M^2)$ can be interpreted as the non-leakage probability ■

Case where the domain V is a parallelepiped of length a, width b and height H.

One can prove that

$$\lambda_0 = (\pi/a)^2 + (\pi/b)^2 + (\pi/H)^2 \tag{9.20}$$

Indeed $\Phi(x, y, z) = \sin\frac{\pi x}{a} \cdot \sin\frac{\pi y}{b} \cdot \sin\frac{\pi z}{H}$ is the associated eigenmode. ■

Orientation

The eigenvalue λ_0 depends only on the domain V occupied by the core. The migration area M^2 cannot be adjusted at will. To obtain $k = 1$, k_∞ must be adjusted, which is possible in a PWR by adjusting the boron concentration in the primary circuit.

The reactivity of the core can also be tuned by using the control rods, but, doing so, the core does not remain homogeneous. The general theory can still be applied but the results are not as simple as in the case of a homogeneous core.

Remark 9.6 *Effectof core dilatation.* We select the origin of our axis system in domain V and we make a dilatation of V in V_θ by the following transform $\{x, y, z\} \to \{x', y', z'\}$ with $x' = \theta x$, $y' = \theta y$ and $z' = \theta z$. We assume $\theta > 1$.

Let Φ be the eigenfunction associated to λ_0. We let

$$\Phi'(x', y', z') = \Phi\left(\frac{x'}{\theta}, \frac{y'}{\theta}, \frac{z'}{\theta}\right)$$

So that

$$\Delta\Phi'\left(\frac{x'}{\theta}, \frac{y'}{\theta}, \frac{z'}{\theta}\right) = \frac{1}{\theta^2}\,\Delta\Phi = -\frac{\lambda_0}{\theta^2}\Phi\left(\frac{x'}{\theta}, \frac{y'}{\theta}, \frac{z'}{\theta}\right)$$

Then, we have

$$\lambda_0' = \frac{\lambda_0}{\theta^2}$$

The greater the domain, the smaller λ_0' so that (9.19) proves that k gets closer and closer to k_∞.

In other words, the lower the leakage.

Now, if $k_\infty > 1$, by decreasing the domain V we can increase the leakage and get $k = 1$.

This is what we call a critical size problem, as we shall see in Chap. 14. ■

Exercise 9.1 The Bessel function J_0 is defined for $x > 0$ as a solution of the ordinary differential equation:

$$J_0'' + \frac{1}{x}J_0' + J_0 = 0$$

It can be shown (see books on special mathematical functions) by induction that

$$J_0(x) = \sum_{j \geq 0} \frac{(-1)^j}{j!j!} \left(\frac{x}{2}\right)^{2j}$$

It is then quite easy to show that $J_0(x) \geq 0$ in the neighborhood of $x = 0$.

Let $x_0 > 0$ denote the smallest solution to the equation $J_0(x) = 0$, we deduce that $J_0(x) \geq 0$. for $0 \leq x \leq x_0$; it is not difficult to calculate $x_0 \cong 2.40483$.

1. Prove that

$$\Phi_0(r) = J_0\left(x_0 \frac{r}{R}\right)$$

   Is the fundamental eigenmode of minus the Laplace operator on the disk of radius R, with the Dirichlet boundary condition at the edge; show that $\mu_0 = \left(\frac{x_0}{R}\right)^2$ is the associated eigenvalue.
2. Deduce that for the fundamental eigenmode of minus the Laplace operator on a cylinder of radius R and height H we have

$$\lambda_0 = \left(\frac{x_0}{R}\right)^2 + \left(\frac{\pi}{H}\right)^2$$

3. We have seen above that the non-leakage coefficient in a homogeneous domain is

$$\mathcal{L} = \frac{1}{1 + \lambda_0 M^2}$$

   where M^2 is the migration area.

We wish to maximize $\mathcal{L}$ for our cylinder and then minimize λ_0 by adjusting parameters H and R but under the constraint

$$\pi H R^2 = V$$

That is at constant volume.
Prove that the choice

$$H = \frac{\pi\sqrt{2}}{x_0} R$$

Is optimal.

Application: The core of a PWR1300 is approximately a cylinder with volume $V = 33\ \text{m}^3$ and height $H = 4$ m. (Note that this choice of parameters is not optimal according to our exercise.)

Compute λ_0 and then, given that $\mathcal{L}_t = 0.98$, compute the migration area M^2. ■ (*solution p. 172*)

1D Example

We assume the core of a nuclear reactor to be a slab bounded by planes $z = 0$ and $z = a$ but infinite along directions x and y. The medium is supposed homogeneous, so that the diffusion coefficient D is constant. The same for the macroscopic cross sections Σ_0, Σ_f and the effective multiplication parameter ν_e; the neutron flux Φ satisfies then the equation

$$-D\Phi'' + \Sigma_0\Phi = \nu_e\Sigma_f\Phi \quad \text{for } 0 \leq z \leq a \tag{9.21}$$

As for the boundary conditions we shall take

$$\Phi'(0) = 0 \quad \text{(symmetry condition)} \tag{9.22}$$

and (to represent an absorbing medium characterized by an adjustable parameter δ):

$$\delta\Phi'(a) + \Phi(a) = 0 \tag{9.23}$$

We assume that $k_\infty = \nu_e \frac{\Sigma_f}{\Sigma_0} > 1$ so that the core can be critical.

Let $\beta = \sqrt{(\nu_e\Sigma_f - \Sigma_0)/D}$ so that we have:

$$-\Phi'' = \beta^2\Phi \quad \text{for } 0 \leq z \leq a$$

and

$$\Phi(z) = A\cos\beta z + B\sin\beta z$$

From (9.22) we conclude that $B = 0$.

From (9.23) we get

$$-\delta\beta\sin\beta a + \cos\beta a = 0$$

$$\text{cotg}\,\beta a = \delta\beta \tag{9.24}$$

We shall then determine the possible intersections between curve $\beta \to \text{cotg}\,\beta a$ and the straight line $\beta \to y = \delta\beta$.

Figure 9.1 represents the situation when $a = \pi$ and $\delta = 2$.

To obtain the general case, just replace 1 by $\frac{\pi}{a}$, 2 by $\frac{2\pi}{a}$, … on the horizontal axis and adjust the slope of the straight line $\beta \to y = \delta\beta$.

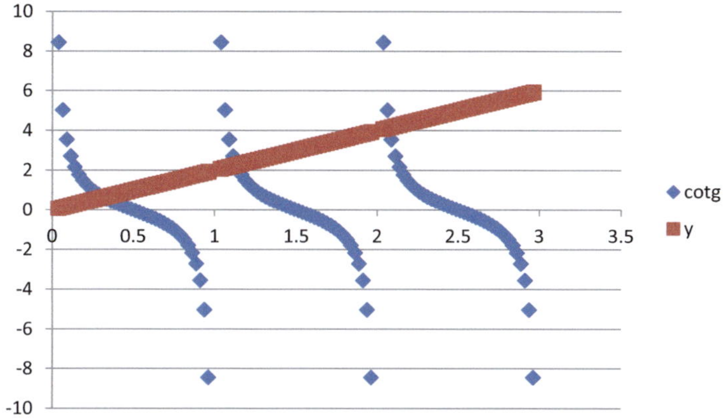

Fig. 9.1 Plot of curves $\beta \to \text{cotg}\, \beta a$ and $\beta \to y = \delta\beta$ (on the axis appear the values of $\beta a/\pi$)

We see that there is a countable set of solutions to (9.23) $\beta_0 < \beta_1 < \beta_2 < \cdots$ but only one solution β_0 in $\left[0, \frac{\pi}{2a}\right]$. We deduce that there is a countable set of eigenmodes $\Phi_i(z) = \cos(\beta_i z)$ but only $\Phi_0(z) = \cos(\beta_0 z)$ is positive in the interval $0 \leq z \leq a$. Then it is the fundamental eigenmode.

Note that in the particular case $\delta = 0$, we obtain $\beta_i = (2i + 1)\frac{\pi}{2a}$ so that $\cos(\beta a) = 0$.

Assume $\beta^2 = (\nu_e \Sigma_f - \Sigma_0)/D < \frac{\pi^2}{4a^2}$ which means that the core is sub-critical with $\delta = 0$, but $(\nu_e \Sigma_f - \Sigma_0)/D > 0$ (i.e. $k_\infty > 1$), we see that adjusting parameter δ to some finite value $\delta > 0$ the core will be critical.

To adjust the absorption of plates located in $z = \pm a$, is a possible way to control the criticality. ■

Exercise 9.2 Prove that $k = k_\infty \frac{\text{absorption}}{\text{absorption+leakage}}$ (solution p. 174). ■

Exercise 9.3 *Effectof absorbing bars*. The H-height core is described by an axial 1D model for $0 \leq z \leq H$.

We let

$$\Sigma_0(z) = \Sigma_B \text{ for the lower part } (0 < z < a)$$

$$\Sigma_0(z) = \Sigma_H \text{ for the higher part } (a < z < H)$$

The problem to be solved is to find $\Phi \geq 0$ such that

$$-D\Phi'' + \Sigma_0(z)\Phi = \nu_e \frac{\Sigma_f}{k}\Phi \text{ for } 0 < z < H \tag{9.25}$$

$$\Phi(0) = 0 \tag{9.26}$$

$$\Phi(H) = 0 \tag{9.27}$$

1. We let:

$$M_B^2 = D/\Sigma_B \quad M_H^2 = D/\Sigma_H$$

$$k_B = \nu_e \frac{\Sigma_f}{\Sigma_B} \quad k_H = \nu_e \frac{\Sigma_f}{\Sigma_H}$$

With the following data:
$H = 400\,\text{cm}$, $D = 11.661\,\text{cm}$, $\Sigma_B = 0.23798\,\text{cm}^{-1}$, $\Sigma_H = 0.24490\,\text{cm}^{-1}$ and $\nu_e \Sigma_f = 0.24\,\text{cm}^{-1}$, compute k for $a = H$ then $a = 0$.
2. We now consider the general case $0 < a < H$. Explain why Φ and its derivative Φ' are continuous in $z = a$.
3. For $k > k_H$, we define $\alpha = \alpha(k)$ and $\beta = \beta(k)$ by

$$\alpha^2 = \frac{k_B/k - 1}{M_B^2}$$

$$\beta^2 = \frac{1 - k_H/k}{M_H^2}$$

so that

$$-\Phi'' = \alpha^2 \Phi \quad \text{for } 0 < z < a \tag{9.28}$$

$$\Phi'' = \beta^2 \Phi \quad \text{for } a < z < H \tag{9.29}$$

Prove that there exist A and B such that

$$\Phi(z) = \begin{Bmatrix} A \sin(\alpha z) & \text{for } 0 < z < a \\ B\,\text{sh}(\beta(H - z)) & \text{for } a < z < H \end{Bmatrix}$$

4. Prove that

$$f(k) = \frac{\alpha(k)}{\text{tg}(\alpha(k)a)} + \frac{\beta(k)}{\text{th}(\beta a)} = 0$$

5. By trials and errors solve equation $f(k) = 0$ for $a = 100$, $a = 200$, $a = 300$.
6. For $a < 70$, the case $k < k_H$ cannot be excluded. Show that, in this case, if we let $\beta = \beta(k)$ such that $\beta^2 = \frac{k_H/k - 1}{M_H^2}$ we have

$$\Phi(z) = \begin{Bmatrix} A\sin(\alpha z) & \text{for } 0 < z < a \\ B\sin(\beta(H - z)) & \text{for } a < z < H \end{Bmatrix}$$

7. Prove that in this case $k < k_{\mathrm{H}}$, we must have

$$g(k) = \frac{\alpha(k)}{\mathrm{tg}(\alpha(k)a)} + \frac{\beta(k)}{\mathrm{tg}(\beta(k)(H - a))} = 0$$

8. By trials and errors solve equation $g(k) = 0$ for $a = 50$.
9. Plot curve $a \to k$. What do you observe?
10. Plot the curve $z \to \Phi(z)$ for $a = 200$. What do you observe? ■ *(solution p. 174)*

Exercise 9.4 (*Reflectors*) We have seen in Exercise 8.2 that if there is a neutron current J_+ going from the core to a slab of light water, we have $J_- = \theta J_+$ with $\theta = 0.7437$.

We have seen in Exercise 8.3 that, for heavy water in the same conditions we also have $J_- = \theta J_+$ but with $\theta = 0.926$.

When the reflector is replaced by vacuum, we have $J_- = 0$ so that $\theta = 0$.

The purpose of this exercise is to evaluate the benefit of a reflector in 1D.

For this, we shall use the one group model:

$$-M^2\Phi'' + \Phi = k_\infty\Phi \quad \text{for } 0 < z < a \tag{9.30}$$

$$\Phi'(0) = 0 \tag{9.31}$$

$$J_-(a) = \theta J_+(a) \tag{9.32}$$

In the above, the height of the core is supposed to be equal to $2a$, so that (9.31) is a symmetry condition. The diffusion coefficient D is assumed to be constant.

1. Prove that $\Phi(z) = \cos(\alpha z)$.
2. Prove that α is a solution of

$$\alpha \tan(\alpha a) = \frac{1 - \theta}{2D(1 + \theta)} \tag{9.33}$$

3. Take $a = 200\,\mathrm{cm}$ and $D = 1$ cm; solve Eq. (4) by trials and errors for $\theta = 0$, $\theta = 0.7437$ and $\theta = 0.926$.
4. Assuming we can take $M^2 = 500\,\mathrm{cm}^2$ evaluate the value of k_∞ that makes the core critical in all 3 cases. ■ *(solution p. 176)*

Remark 9.7 In Exercise 9.4, the migration area M^2 has been evaluated by taking into account the fact there are fast neutrons in the core (see further Remark 10.2).

In a PWR the non-leakage probability is well known (see Fig. 5.1). Of course, the non-leakage probability is higher in a 1D geometry like here. Since the core is critical we have $k = 1$ so that $\mathcal{L} = 1/k_\infty$. The exercise shows the benefit induced by a reflector in 1D. For a cylinder the benefit is obviously higher with a reflector both on the flat faces and on the lateral faces of the cylinder. ■

Chapter 10
Two Groups Diffusion Model

In this chapter we shall study the two groups diffusion model that is more elaborate, while remaining relatively simple. In such a model the first group is for fast neutrons and the second one for thermal neutrons. It is useful insofar as neutron cross sections are much larger for thermal neutrons than for fast neutrons. We shall introduce migration areas for each group. They have not the same order of magnitude. This new model is useful to take into account fast fissions in group 1 and poisoning in group 2.

The two groups diffusion model will be used in Chap. 17.

We will now study the following model for a homogeneous core:

$$-D_1 \Delta\psi + \Sigma_1 \psi = \varepsilon\nu\Sigma_f \varphi \quad \text{for } x \in V \tag{10.1}$$

$$-D_2 \Delta\varphi + \Sigma_2 \varphi = p\Sigma_1 \psi \quad \text{for } x \in V \tag{10.2}$$

$$\varphi = \psi = 0 \quad \text{on the boundary } \Gamma \tag{10.3}$$

This is a system of two coupled diffusion equations whose solution $\{\varphi, \psi\}$ comprises two neutron fluxes: ψ is the fast neutron flux and φ the thermal neutron flux.

To keep the calculations simple, we have neglected the extrapolation length and adopted a simpler, slightly more absorbing boundary condition (so called Dirichlet boundary condition).

On the right hand side of Eq. (10.1), a source term appears to model the fast neutrons issued from Uranium fission.

On the right hand side of Eq. (10.2) we see the appearance of a source term that models thermal neutrons that were initially fast and survived after slowing down.

Σ_2 is therefore the macroscopic absorption cross-section of the core in the thermal domain.

B. Mercier, *Simple Models for Understanding Nuclear Reactor Physics*,
https://doi.org/10.1007/978-3-031-98532-4_10

Σ_1 is the (macroscopic) fast-group disappearance cross-section for a fast neutron. Notice that a fraction p of them survive and appear as a source term in (10.2).

To find the critical equation associated with this problem, which is (without appearing to be) an eigenvalue problem, we shall introduce the fundamental eigenmode Φ_0 satisfying:

$$\begin{aligned} -\Delta\Phi_0 &= \lambda_0\Phi_0 \quad \text{in } V \\ \Phi_0 &\geq 0 \quad \text{in } V \\ \Phi_0 &= 0 \quad \text{on } \Gamma \end{aligned} \tag{10.4}$$

and we shall look for $\{\varphi, \psi\}$ solution of (10.1), (10.2) and (10.3) in the form

$$\{\varphi, \psi\} = \{\alpha\Phi_0, \beta\Phi_0\}$$

where α and $\beta \in \boldsymbol{R}$.

Introducing $\varphi = \alpha\Phi_0$ and $\psi = \beta\Phi_0$ in (10.1) and (10.2) we get

$$\begin{aligned} &-\beta D_1\Delta\Phi_0 + \Sigma_1\beta\Phi_0 = \varepsilon\nu\Sigma_f\alpha\Phi_0 \\ &-\alpha D_2\Delta\Phi_0 + \Sigma_2\alpha\Phi_0 = p\Sigma_1\beta\Phi_0 \end{aligned}$$

So that by using (10.4) and simplification by Φ_0:

$$\begin{pmatrix} \varepsilon\nu\Sigma_f & -(\Sigma_1 + D_1\lambda_0) \\ -(\Sigma_2 + D_2\lambda_0) & p\Sigma_1 \end{pmatrix} \begin{pmatrix} \alpha \\ \beta \end{pmatrix} = \begin{pmatrix} 0 \\ 0 \end{pmatrix}$$

consequently, the determinant of this 2×2 matrix must vanish, hence

$$\varepsilon p\nu\Sigma_f\Sigma_1 - (\Sigma_1 + D_1\lambda_0)(\Sigma_2 + D_2\lambda_0) = 0$$

Dividing by $\Sigma_1\Sigma_2$ we get

$$\varepsilon p\nu\frac{\Sigma_f}{\Sigma_2} = \left(1 + \frac{D_1}{\Sigma_1}\lambda_0\right)\left(1 + \frac{D_2}{\Sigma_2}\lambda_0\right)$$

That is our critical equation.

Indeed

$$\varepsilon p\nu\frac{\Sigma_f}{\Sigma_U}\frac{\Sigma_U}{\Sigma_2} = \varepsilon p f\eta = k_\infty = \left(1 + \frac{D_1}{\Sigma_1}\lambda_0\right)\left(1 + \frac{D_2}{\Sigma_2}\lambda_0\right)$$

and if we let

$$k = \frac{k_\infty}{\left(1 + \frac{D_1}{\Sigma_1}\lambda_0\right)\left(1 + \frac{D_2}{\Sigma_2}\lambda_0\right)} \tag{10.5}$$

Our critical equation just gives $k = 1$ then we see that the k we obtain is the effective multiplication factor.

In analogy with the one-group diffusion model, we introduce the following migration areas

$M_1^2 = \frac{D_1}{\Sigma_1}$ and $M_2^2 = \frac{D_2}{\Sigma_2}$.

One in the fast neutrons group. The other for thermal neutrons.

We get the 6 factors formula:

$$k = \varepsilon \cdot \mathscr{L}_f \cdot p \cdot \mathscr{L}_t \cdot f \cdot \eta$$

with

$$\mathscr{L}_f = \frac{1}{1 + M_1^2 \lambda_0} \tag{10.6}$$

That is the non-leakage probability in the fast neutrons group and

$$\mathscr{L}_t = \frac{1}{1 + M_2^2 \lambda_0} \tag{10.7}$$

That is the non-leakage probability in the thermal neutrons group.

Remark 10.1 The one-group model gives

$$k = \frac{k_\infty}{1 + \lambda_0 M^2}$$

This is close to the result given by our two groups model provided that

$$M^2 = M_1^2 + M_2^2$$

Indeed, the core of a PWR is quite large, then λ_0 is small and even $\lambda_0 M^2$ is small. The non-leakage probabilities $\mathscr{L}_f$ et $\mathscr{L}_t$ are then close to 1. So, we have

$$\frac{1}{1 + \lambda_0 M^2} \cong \frac{1}{1 + M_1^2 \lambda_0} \frac{1}{1 + M_2^2 \lambda_0}$$

Which means that in the one group model it is better to take $M^2 = M_1^2 + M_2^2$ than $M^2 = M_2^2$. ■

Exercise 10.1 (*Another 2 Groups Model*) Let Ψ (resp. Φ) denote the fast (resp. thermal) neutron flux in the core, assumed homogeneous.

This new model also involves two coupled diffusion equations

$$-D_1 \frac{\partial^2}{\partial z^2} \Psi + \Sigma_1 \Psi = \nu \Sigma_{f2} \Phi + \nu \Sigma_{f1} \Psi$$

$$-D_2\frac{\partial^2}{\partial z^2}\Phi + \Sigma_2\Phi + \sigma X\Phi = p\Sigma_1\Psi$$

where

- D_1 (resp. D_2) denotes the diffusion coefficient for the fast (resp. thermal) neutrons group
- Σ_{f1} (resp. Σ_{f2}) denotes the macroscopic fission cross section in the fast (resp. thermal) neutrons group
- ν is the average number of (fast) neutrons generated by each fission (for the sake of simplicity we assume it is the same both for thermal and fast fissions)
- Σ_1 denotes the (macroscopic) fast-group disappearance cross-section for a fast neutron
- p denotes the resonance escape probability, so that a proportion $(1-p)$ of the neutrons that disappear from the fast group are physically absorbed, and a proportion p of them reappear in the thermal group.
- Σ_2 denotes the (macroscopic) absorption cross section in the thermal group.
- σ denotes the microscopic capture cross section of a neutronic poison (e.g.: boron) for thermal neutrons (assumed to be negligible for fast neutrons),
- X is the number of atoms of neutronic poison per cm^3 of core.

This system of two equations is complemented by boundary conditions in $z = 0$ and in $z = H$, where H denotes the height of the core. (Its radius is supposed to be much larger that its height.)

We select the Dirichlet boundary conditions

$$\Psi(0) = \Psi(\mathrm{H}) = \Phi(0) = \Phi(H) = 0$$

We let $M_1^2 = D_1/\Sigma_1$, $M_2^2 = D_2/\Sigma_2$, $k_{\infty,1} = \nu\Sigma_{f1}/\Sigma_1$, $k_{\infty,2} = \nu p\Sigma_{f2}/\Sigma_2$, $Q = \Sigma_2/(p\Sigma_1)$, $P = \sigma X/\Sigma_2$.

So that

$$-M_1^2\frac{\partial^2}{\partial z^2}\Psi + \left(1 - k_{\infty,1}\right)\Psi = Qk_{\infty,2}\Phi \tag{10.8}$$

$$-M_2^2\frac{\partial^2}{\partial z^2}\Phi + (1+P)\Phi = Q^{-1}\Psi \tag{10.9}$$

1. We assume X does not depend on z; prove that if $\frac{k_{\infty,2}}{1+P} + k_{\infty,1} > 1$

$$\Psi(\mathrm{z}) = \Psi_0\sin\left(\frac{\pi z}{H}\right)$$
$$\Phi(\mathrm{z}) = \Phi_0\sin\left(\frac{\pi z}{H}\right)$$

where Ψ_0 are Φ_0 some constant numbers.

2. Let $\lambda_0 = \pi^2/H^2$ prove that

$$\begin{pmatrix} 1+\lambda_0 M_1^2 - k_{\infty,1} & -Qk_{\infty,2} \\ -Q^{-1} & 1+P+\lambda_0 M_2^2 \end{pmatrix} \begin{pmatrix} \Psi_0 \\ \Phi_0 \end{pmatrix} = \begin{pmatrix} 0 \\ 0 \end{pmatrix}$$

3. Deduce that

$$\left(1+\lambda_0 M_1^2 - k_{\infty,1}\right)\left(1+P+\lambda_0 M_2^2\right) = k_{\infty,2} \tag{10.10}$$

   Explain why this is the critical equation.
4. Prove that if $P > 0$

$$\frac{k_{\infty,1}}{1+\lambda_0 M_1^2} + \frac{k_{\infty,2}}{\left(1+\lambda_0 M_1^2\right)\left(1+\lambda_0 M_2^2\right)} > 1 \tag{10.11}$$

   Conversely, if (10.12) is satisfied then (10.11) has one solution $P > 0$.
5. Prove that

$$k_{\infty,2} = p\eta f$$

   where η and f are 2 factors appearing in Fermi's 4 factors formula.
6. Numerical application: take.

   - $\Sigma_{f1} = 0.0005\,\text{cm}^{-1}$
   - $\Sigma_{f2} = 0.081585\,\text{cm}^{-1}$
   - $\nu = 2.42$
   - $\Sigma_1 = 0.025\,\text{cm}^{-1}$
   - $p = 0.616$
   - $\Sigma_2 = 0.119\,\text{cm}^{-1}$.

   Compute $k_{\infty,1}$ then $k_{\infty,2}$.
7. Take $H = 400\,\text{cm}$, $M_1^2 = 50\,\text{cm}^2$ and $M_2^2 = 9\,\text{cm}^2$ compute P.
8. Compute Ψ_0/Φ_0.
9. Compute the fast fission factor ε (*solution*, p. 177). ■

Remark 10.2 Set $\psi = Q^{-1}\Psi$; from (10.8) and (10.9) we deduce (with $P = 0$)

$$-M_1^2 \frac{\partial^2}{\partial z^2}\psi + \left(1 - k_{\infty,1}\right)\psi = k_{\infty,2}\Phi \tag{10.12}$$

$$-M_2^2 \frac{\partial^2}{\partial z^2}\Phi + \Phi = \psi \tag{10.13}$$

Let us assume that $M_2^2 \ll M_1^2$.
We do not assume any more that

$$\psi(0) = \psi(\mathrm{H}) = \Phi(0) = \Phi(H) = 0$$

Rather, we assume that they vanish at the extrapolation distance that may not be the same for ψ and Φ.

Since both ψ and Φ follow approximately function $z \to \sin \pi z/H$ then

$$-M_2^2 \frac{\partial^2}{\partial z^2}\Phi \sim M_2^2\left(\frac{\pi}{H}\right)^2 \Phi$$

Provided that $M_2^2\left(\frac{\pi}{H}\right)^2 \ll 1$ then $\Phi \cong \psi$.

Reporting in (10.13) we see that

$$-M_1^2 \frac{\partial^2}{\partial z^2}\psi + \psi = \left(k_{\infty,1} + k_{\infty,2}\right)\psi$$

So it is the extrapolation distance to be used for ψ that matters, and, for ψ, we just have to solve a one group eigenvalue problem with M_1^2 as the migration area. This will be illustrated further in §17.2.

So, in such a situation, what matters is the migration area for the group of fast neutrons. ■

Chapter 11
Delayed Neutrons and Reactor Kinetics

The fact that the neutron flux is solution of an eigenvalue problem seems to indicate that the reactor is unstable. Either $k > 1$ and the neutron flux increases exponentially. Either $k < 1$ and the neutron flux decreases exponentially! Yes, but what is the time constant of such exponential functions?

Fortunately not all fission neutrons are generated instantaneously: a fraction of them are delayed neutrons and it changes reactor kinetics significantly.

The neutron flux Φ is actually depending on 7 variables: $\Phi = \Phi\left(x, \vec{\Omega}, E, t\right)$.

To accurately compute such a function of 7 variables is a numerical challenge.

In this book, we have studied how Φ depends on E and then on x and $\vec{\Omega}$.

The target of this chapter is to study how it depends on t.

In Chap. 9, Remark 9.3, we have seen that the one group model could be obtained as the limit as $t \to \infty$ of some unsteady diffusion model. However this is not physically relevant since it ignores the *delayed neutrons* phenomenon.

In fact, most fission products are β-emittors; some of them produce, after β-decay, metastable nuclei that will release one neutron. Since β-decay is not instantaneous, such neutrons are called *delayed neutrons.*

Here are two examples of such fission products:

$^{87}\mathrm{Br} \to {}^{87}\mathrm{Kr}^* \to {}^{86}\mathrm{Kr} + n$ (radioactive decay period: 54.5 s)
$^{137}\mathrm{I} \to {}^{137}\mathrm{Xe}^* \to {}^{136}\mathrm{Xe} + n$ (radioactive decay period: 21.8 s).

The nuclei $^{87}\mathrm{Br}$ and $^{137}\mathrm{I}$ are (logically) called "precursors".

By opposition to delayed neutrons, so-called *prompt neutrons* are emitted instantaneously.

More generally (cf. [2, 3]), 6 groups of delayed neutrons are classically considered.

Overall, for every 100,000 neutrons emitted by fission, column 2 of Table 11.1 shows how many are emitted in each of these 6 groups. So, in proportion, there are $(1 - \beta)$ prompt neutrons and β delayed neutrons. Overall, there are less than 0.7% delayed neutrons, but this is enough, as we shall see, to make the reactor controllable

B. Mercier, *Simple Models for Understanding Nuclear Reactor Physics*,
https://doi.org/10.1007/978-3-031-98532-4_11

Table 11.1 6 groups of delayed neutrons

Group	β (pcm)	td (s)	T (s)
1	24	54.5	78.6
2	123	21.8	31.5
3	117	6	8.6
4	262	2.2	3.2
5	108	0.5	0.7
6	45	0.18	0.26
	$\Sigma = 679$	Ave = 7.8	Ave = 11.3

as long as β is not too small. (Column 3 shows the half life and column 4 the time constant of the beta decay)

Neutrons Lifetime

It is the average time for a neutron between its creation and its absorption. We denote it by ℓ.

We have seen in Chap. 4 (Fig. 4.3) that, after it has been created, a neutron performs a random walk in the core. The parameter ℓ is then the average duration of this random walk. It is understandable that it may be greater in a thermal neutron reactor than in a fast neutron reactor.

It is also conceivable that the lifetime may be shorter if the effective multiplication coefficient $k < 1$ and longer if $k > 1$.

The following relation is admitted (see [2])

$$\ell = k \cdot \ell_0 \tag{11.1}$$

where ℓ_0 is the lifetime obtained when the reactor is critical ($k = 1$). For a PWR, the lifetime is typically of the order of 25 microseconds.

In fact, for a neutron, a simple calculation shows that the slowing down phase is quite short ($\sim$ 2 μs). When the neutron enters the thermal domain where its average speed is $v \sim 2200\,\mathrm{m/s}$ and $\Sigma_0 = 0.2\,\mathrm{cm}^{-1}$ its mean free path is $1/\Sigma_0$ so that the neutron remains (in average) a time $1/(v\Sigma_0)$ (that is 22.7 μs) in the thermal domain before being absorbed.

On the other hand we have $k_\infty = \nu\Sigma_f/\Sigma_0$ which shows that Σ_0 is proportional to $(k_\infty)^{-1}$ then to k^{-1}. So (11.1) holds if we neglect the slowing down time.

Now, from the definition of the effective multiplication factor, during time ℓ the number of neutrons in the core increases or decreases in the following way:

$$n(t + \ell) = k \cdot n(t)$$

Then, introducing the *reactivity* $\rho = (k - 1)/k$, we have (we assume ρ does not depend on time)

$$\frac{n(t+\ell)-n(t)}{\ell}=\frac{k-1}{k\cdot\ell_0}n(t)=\frac{\rho}{\ell_0}n(t)$$

which corresponds to the ordinary differential equation

$$\frac{\mathrm{d}n}{\mathrm{d}t}=\frac{\rho}{\ell_0}\cdot n$$

The solution of which is

$$n(t)=n(0)\exp\left(\frac{\rho}{\ell_0}t\right)$$

Note that ρ being small, for $t=\ell$, we have

$$\begin{aligned} n(\ell) &= n(0)\exp(\rho\ell/\ell_0)\cong n(0)(1+\rho\ell/\ell_0) \\ &= n(0)(1+k\rho)=k\cdot n(0) \end{aligned}$$

As an example, when $\rho=100$ pcm (that is $\rho=0.001$) and $t=0.1$ s we would have

$$n(t)\cong 148\,n(0)\quad !!!$$

which means that, without delayed neutrons, the reactor would be dangerously unstable.

The case with delayed neutrons is governed not by a differential equation but by the following differential system:

$$\frac{\mathrm{d}n}{\mathrm{d}t}=\frac{\rho-\beta}{\ell_0}n+\sum_i\lambda_i c_i \tag{11.2}$$

$$\frac{\mathrm{d}c_i}{\mathrm{d}t}=\frac{\beta_i}{\ell_0}n-\lambda_i c_i\quad 1\le i\le 6 \tag{11.3}$$

where we see that a fraction of the increase in the number of neutrons comes from the decay of precursors, and that conversely, fissions feed the precursor population.

The reasoning behind these equations is as follows: if we start with $n(0)$ neutrons, at the next generation, i.e., at time ℓ, we will have:

$$n(t+\ell)=k(1-\beta)n(t)+\ell\left(\sum_i\lambda_i c_i\right)$$

where the first term corresponds to the number of prompt neutrons emitted by fission and the second one to the number of delayed neutrons emitted by precursors during time ℓ.

Equivalently, we have

$$n(t+\ell) - n(t) = -n(t) + k(1-\beta)n(t) + \ell\left(\sum_i \lambda_i c_i\right)$$

Dividing by ℓ and applying $\ell = k\,\ell_0$ we find

$$\frac{n(t+\ell) - n(t)}{\ell} = \frac{\rho - \beta}{\ell_0} n(t) + \left(\sum_i \lambda_i c_i\right)$$

which corresponds to (11.2).

In what follows, we shall replace ℓ_0 by ℓ to somewhat simplify the equations; (11.2) and (11.3) is a differential system with constant coefficients of the form:

$$\frac{du}{dt} = Au$$

where A is the 7×7 matrix: $\left(\begin{bmatrix} \frac{\rho-\beta}{\ell} & \cdots & \lambda_6 \\ \vdots & \ddots & \vdots \\ \frac{\beta_6}{\ell} & \cdots & -\lambda_6 \end{bmatrix}\right)$.

Mathematical Result

If the matrix A is diagonalizable, the general solution is a combination of the 7 particular solutions that are the $e^{\omega_i t}$, where ω_i, $1 \le i \le 7$ are the 7 eigenvalues of matrix A.

The 7 eigenvectors are, of course, solutions of the matrix equation:

$$AU = \omega U$$

That is

$$\omega n = \frac{\rho - \beta}{\ell} n + \sum_i \lambda_i c_i \tag{11.4}$$

$$\omega c_i = \frac{\beta_i}{\ell} n - \lambda_i c_i \tag{11.5}$$

To solve this system of 7 equations with 7 unknowns, we remark that

$$(\omega + \lambda_i)c_i = \frac{\beta_i}{\ell} n$$

Which allows to eliminate c_i from (11.4)

$$\omega n = \frac{\rho - \beta}{\ell} n + \sum_i \frac{\lambda_i}{\omega + \lambda_i} \frac{\beta_i}{\ell} n;$$

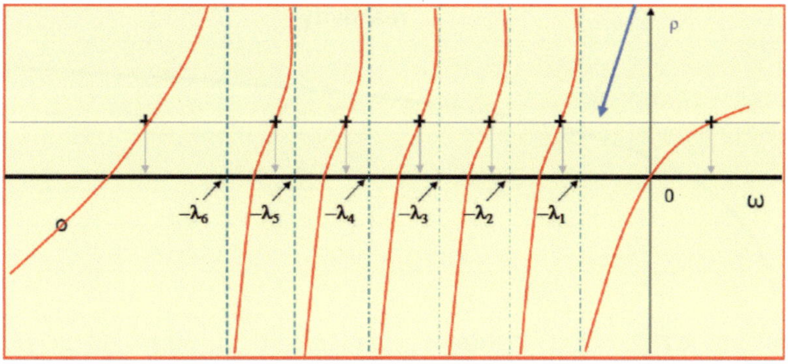

Fig. 11.1 Shape of function $\omega \to f(\omega)$ involved in Nordheim's equation (11.6).

as an eigenvector is necessarily non-zero, we know that $n \neq 0$ and we can therefore simplify by n:

$$\omega = \frac{\rho - \beta}{\ell} + \sum_i \frac{\lambda_i}{\omega + \lambda_i} \frac{\beta_i}{\ell}$$

Hence, using

$$\beta = \sum_i \beta_i$$

$$\omega = \frac{\rho}{\ell} + \sum_i \left(\frac{\lambda_i}{\omega + \lambda_i} - \frac{\omega + \lambda_i}{\omega + \lambda_i} \right) \frac{\beta_i}{\ell} = \frac{\rho}{\ell} - \sum_i \frac{\omega}{\omega + \lambda_i} \frac{\beta_i}{\ell}$$

This gives the so-called Nordheim equation [2, 3]:

$$\rho = \ell\omega + \sum_i \frac{\omega}{\omega + \lambda_i} \beta_i \tag{11.6}$$

This equation is equivalent to an algebraic equation of degree 7, which has 7 roots ω_i, $1 \leq i \leq 7$.

It's an equation that's best solved graphically.

Let

$$f(\omega) = \ell\omega + \sum_i \frac{\omega}{\omega + \lambda_i} \beta_i$$

We notice that function $\omega \to f(\omega)$ has 6 poles $\omega = -\lambda_i$; this guarantees the existence of 7 distinct real roots $\omega_7 < \omega_6 < \omega_5 < \omega_4 < \omega_3 < \omega_2 < \omega_1$.

Figure 11.1, which shows the shape of the curve $\omega \to f(\omega)$, proves that ω_1 has the same sign as ρ.

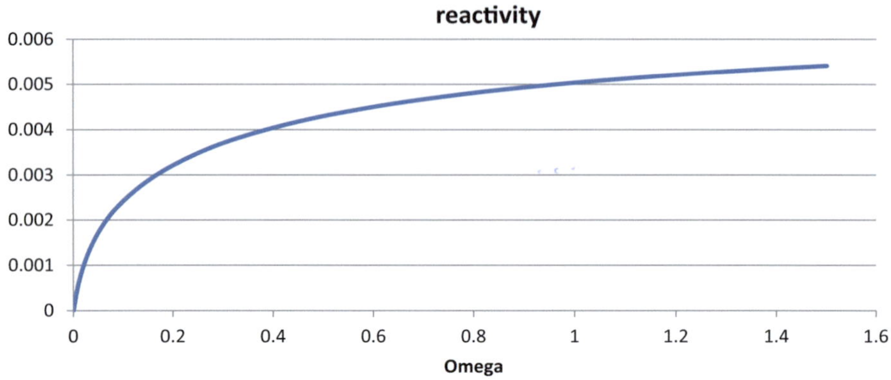

Fig. 11.2 Zoom of the previous curve for $\omega \geq 0$

Figure 11.1, illustrates the case $\rho > 0$; it is reproduced from [10].

Since the general solution of (11.2) and (11.3) is a combination of the 7 particular solutions that are the $e^{\omega_i t}$, we see that $e^{\omega_1 t}$ is the dominating term, particularly if $\omega_1 \geq 0$. In fact, all the other solutions $e^{\omega_i t}$ for $2 \leq i \leq 7$ are exponentially decreasing when $t \to \infty$.

If we perform a zoom on the right part $\omega \geq 0$ of this curve (cf. Fig. 11.2) we see that for a 100 pcm reactivity ($\rho = 0.001$) we obtain $\omega = 0.02\ \mathrm{s}^{-1}$ in such a way that $n(t) \approx e^{0.02\,t}$ so that within 10 s the reactor power will have increased by only 22%, which is very reasonable and perfectly compatible with control requirements.

Remark 11.1 We see on Fig. 11.2 that if ρ is small, ω is small; more precisely

$$f(\omega) \cong f'(0) \cdot \omega = \left(\ell + \sum_i \beta_i/\lambda_i\right) \cdot \omega = \overline{\ell} \cdot \omega$$

so that, noting that $f(\omega) = \rho$,

$$\omega \cong \frac{\rho}{\overline{\ell}}.$$

It is as if the neutron lifetime ℓ *were replaced by* $\overline{\ell} = \ell + \sum_i \beta_i/\lambda_i$. ■

On the contrary, if $\rho > \beta$, from (11.2) we get that

$$\frac{\mathrm{d}n}{\mathrm{d}t} - \frac{\rho - \beta}{\ell} n > 0$$

so that

$$n(t) > n(0) \exp\left(\frac{\rho - \beta}{\ell} t\right),$$

Table 11.2 Delayed neutrons fraction for different fissile nuclei

Fissile nucleus	β (pcm)
Thorium 232 (F)	2433
Uranium 233	296
Uranium 235	679
Uranium 238 (F)	1828
Plutonium 239	224
Plutonium 240 (F)	292
Plutonium 241	535

which means that the prompt neutrons are sufficient to make the core supercritical so that the core power increases very fast. We say that there is *prompt criticality*.

Practically, the reactor's GOR (General Operating Rules) require to limit reactivity to $\rho < \beta/2$ that is $\rho \cong 0.003$, or 300 pcm.

Figure 11.2 shows that, with such a limitation we have $\omega_1 < 0.16\,\mathrm{s}^{-1}$ which is reasonable.

Note that parameter β depends on the fissile material used in the core.

Table 11.2 shows that β is significantly smaller for ^{239}Pu than for ^{235}U. This explains why the Operating Rules will be more stringent for a PWR reactor using MOX fuel. Even for a reactor using UOX fuel, parameter β will be smaller at EOC (End of Cycle) that at BOC (Beginning of Cycle) due to fuel evolution (see Chap. 13).

Exercise 11.1 Consider the point kinetics model, with a single group of delayed neutrons. We introduce:

ρ reactivity $\left(\rho = \frac{k-1}{k}\right)$,
β fraction of delayed neutrons,
λ the precursor decay constant,
ℓ neutron lifetime in the critical reactor,
$n = n(t)$ number of neutrons in the core at time t,
$c = c(t)$ number of precursors in the core at time t.

We know that we end up with the system of equations

$$\frac{\mathrm{d}n}{\mathrm{d}t} = \frac{\rho - \beta}{\ell} n + \lambda c \tag{11.7}$$

$$\frac{\mathrm{d}c}{\mathrm{d}t} = \frac{\beta}{\ell} n - \lambda c$$

However, for the purpose of this exercise, it will be assumed that there is a permanent source $S > 0$ of neutrons in the core, so Eq. (11.7) is replaced by

$$\frac{\mathrm{d}n}{\mathrm{d}t} = \frac{\rho - \beta}{\ell} n + \lambda c + S \tag{11.8}$$

1. We first assume the core to be critical ($\rho = 0$). Prove that,

$$\frac{d^2n}{dt^2} + \left(\lambda + \frac{\beta}{\ell}\right)\frac{dn}{dt} = \lambda S. \tag{11.9}$$

2. Prove that (11.9) has a general solution of the following form:

$$n(t) = A + B\exp\left(-\left(\lambda + \frac{\beta}{\ell}\right)t\right) + \frac{\lambda S}{\lambda + \frac{\beta}{\ell}}t$$

where A and B are some constants. Deduce that $n(t) \to \infty$ when $t \to \infty$.

3. Let

$$A = \begin{pmatrix} \frac{\rho-\beta}{\ell} & \lambda \\ \frac{\beta}{\ell} & -\lambda \end{pmatrix} \text{ and } f = \begin{pmatrix} S \\ 0 \end{pmatrix}$$

Prove that if $\rho \neq 0$ the linear system

$$Au_1 = -f$$

Has a unique solution $u_1 = \begin{pmatrix} n_1 \\ c_1 \end{pmatrix}$.

4. Let

$$u(t) = \begin{pmatrix} n(t) - n_1 \\ c(t) - c_1 \end{pmatrix}$$

Prove that.

$$\frac{du}{dt} = Au.$$

5. Deduce that, if $\rho < 0$ then $u(t) \to 0$ when $t \to \infty$, and then

$$n(t) \to n_1$$

$$c(t) \to c_1$$

6. Compute n_1 as a function of S, ℓ and ρ; deduce that when $S > 0$ it is possible to obtain n_1 as big as we wish by selecting $\rho < 0$ possibly close to 0. ■ (*solution p. 180*)

SubcriticalApproach: When solving Exercise 11.1 we obtain

$$n_1 = -\frac{\ell}{\rho}S \tag{11.10}$$

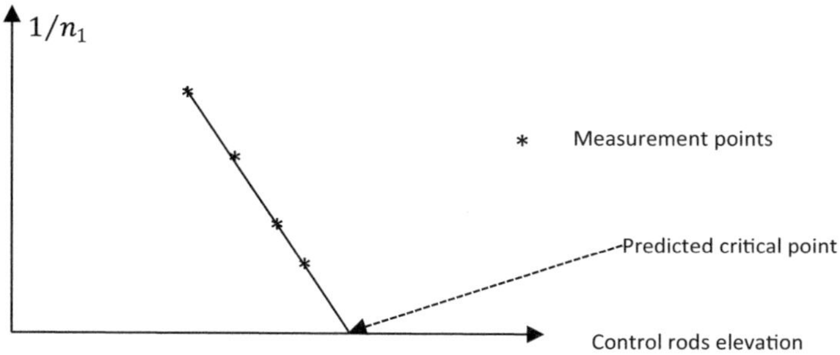

Fig. 11.3 Subcritical approach

so that if we make $\rho \to 0$, with $\rho < 0$ we see that $n_1 \to \infty$.

In practice, to start a reactor with a new core, you need a start-up source, represented here by S.

By using the control bars, the operator can adjust reactivity.

The control bars elevation is then a possible indicator of reactivity. However, the operator would like to know at which elevation he/she will get $\rho = 0$.

When $S > 0$, for a given $\rho < 0$ the operator is able to measure n_1 with the neutron detectors.

On a graph (see Fig. 11.3), he is able to plot the control rods elevation on the x-axis and $\frac{1}{n_1}$ on the y-axis. Formula (11.10) shows that the curve he obtains is a straight line. By extending the curve toward the horizontal axis, he can deduce the control bars elevation at criticality.

Note that the control bars elevation can be replaced by the Boron concentration.

To achieve the core divergence, the operator will adjust either the control bars or the boron concentration to criticality, so that, as we have seen in question 2 of the exercise, $n(t) \to \infty$. When the reactor power has reached the desired level, the operator will remove the start-up source. ■

Exercise 11.2 (*Simplified reactor kinetics*) We still use the point kinetics model, with a single group of delayed neutrons, as in the previous exercise, but we assume $S = 0$, so that we have

$$\frac{\mathrm{d}n}{\mathrm{d}t} = \frac{\rho - \beta}{\ell} n + \lambda c \tag{11.11}$$

$$\frac{\mathrm{d}c}{\mathrm{d}t} = \frac{\beta}{\ell} n - \lambda c \tag{11.12}$$

1. We denote by ω_1 and ω_2 the eigenvalues of A, where A is the 2×2 matrix

$$\begin{pmatrix} \frac{\rho-\beta}{\ell} & \lambda \\ \frac{\beta}{\ell} & -\lambda \end{pmatrix}$$

Prove that if $\rho < 0$ both eigenvalues are negative, and that if $\rho > 0$, one of them (let's call it ω_1) is positive. Conclude that the system is critical if $\rho = 0$.

2. What happens when $\rho > \beta$?
3. We now assume that $\rho > 0$ but $\rho - \beta < 0$

   Prove that, if λ is small compared to both $\frac{\rho}{\ell}$ and $\frac{\beta-\rho}{\ell}$ then

$$\omega_2 \cong \lambda \frac{\rho}{\beta - \rho}$$

   How should we limit ρ for the reactor to be controllable?
4. Numerical application: let $\beta = 380\text{pcm}$, $\rho = 190\text{pcm}$, $\ell = 2.5 \times 10^{-5}$ s, $\lambda = 0.09\,\text{s}^{-1}$. Precisely compute ω_2 then compute it by using the above approximation.

   (NB. We check that $\lambda\ell = 2.25$ pcm so that $\rho < \beta - \lambda\ell$.)
5. Explain why $\frac{\rho-\beta}{\ell} - \lambda - \lambda\frac{\rho}{\beta-\rho}$ is a good approximation to ω_1.
6. Suppose we start from a critical state $\{n_0, c_0\}$ such that

$$\frac{\beta}{\ell} n_0 - \lambda c_0 = 0$$

At time $t = 0$, we make a reactivity step: reactivity jumps from 0 to $\rho > 0$.

Prove that $n(t)$ will rapidly increase from n_0 to $A_2 = \frac{\beta}{\beta-\rho} n_0$ (this is called the "prompt jump") and then (more slowly) increase like $\frac{\beta}{\beta-\rho} n_0 \exp\left(\lambda \frac{\rho}{\beta-\rho} t\right)$. ■ (*solution p. 183*)

Time Discretization

For PWR system simulation, as we shall see in Chap. 16, the linear differential system (11.11) and (11.12) will have to be replaced by a nonlinear system. Therefore, we shall study the effect of time discretization in the linear case considered in this chapter.

Now, since $|\omega_1| >> \omega_2$ we have two very different time scales (the ratio is close here to 850), and (11.11) and (11.12) is called a stiff system of differential equations. (see Wikipedia)

"*Stiff equations are equations where certain implicit methods, in particular backward difference scheme (BDF), perform better, usually tremendously better, than explicit ones.*"

To take into account the very short lifetime of prompt neutrons, we shall introduce an implicit scheme coupling the discretized analog to Eqs. (11.11) and (11.12). We introduce a timestep τ; given the values of n (replaced by x in the following) and c at time t, we denote the new values at time $t + \tau$ by x^* and c^*, we require that

$$\frac{x^* - x}{\tau} = \frac{\rho - \beta}{\ell} x^* + \lambda c^*$$

$$\frac{c^* - c}{\tau} = \frac{\beta}{\ell} x^* - \lambda c^*$$

That is $x^*\left(1-\frac{\rho-\beta}{\ell}\tau\right)=\lambda c^*\tau+x$

$$c^*(1+\lambda\tau)=c+\frac{\beta}{\ell}\tau x^*$$

which we can write

$$\begin{pmatrix} A & B \\ C & D \end{pmatrix}\begin{pmatrix} x^* \\ c^* \end{pmatrix}=\begin{pmatrix} x \\ c \end{pmatrix}$$

where

$$A=1-\frac{\rho-\beta}{\ell}\tau \quad B=-\lambda\tau$$

$$C=-\frac{\beta}{\ell}\tau \quad D=1+\lambda\tau$$

Let $\Delta=AD-BC$ we have

$$\begin{pmatrix} x^* \\ c^* \end{pmatrix}=\frac{1}{\Delta}\begin{pmatrix} D & -B \\ -C & A \end{pmatrix}\begin{pmatrix} x \\ c \end{pmatrix} \tag{11.13}$$

On Fig. 11.4, we show what we get with the data of Exercise 11.2 and $n_0=x_0=1$: we compare $\tau=0.05$ s with $\tau=0.1$s.

We can see that the prompt jump defined in the previous exercise (here from n_0 to $2n_0$) is obtained in 1 or 2 steps.

The Backward Difference Scheme appears to be adequate, even with timesteps much larger that the neutron lifetime.

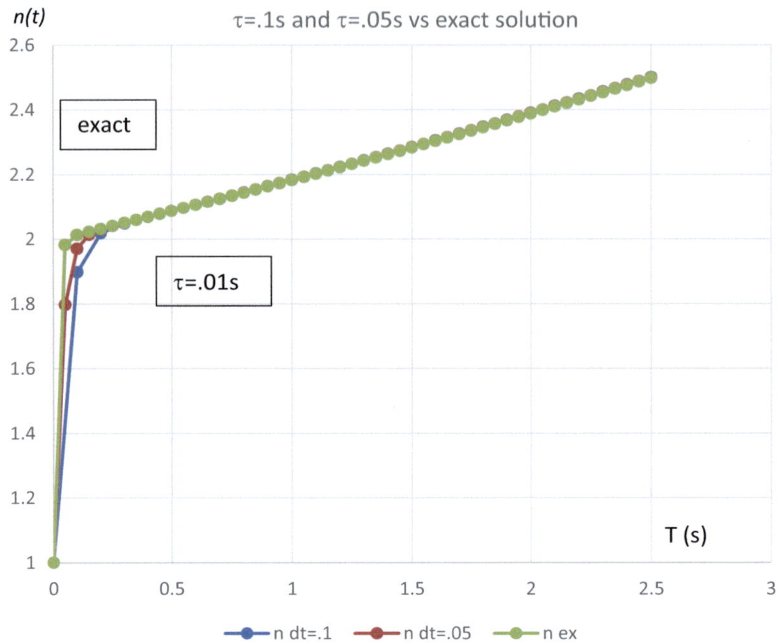

Fig. 11.4 Finite difference solution (BDF) versus exact solution

Chapter 12
Fuel Poisoning by Fission Products

Fuel poisoning by fission products is a significant effect that the operators of a nuclear reactor should properly understand. In this chapter we evaluate the negative reactivity induced by such a poisoning and the related kinetics.

Here is what Nigon [10] says.

"Each fission produces unstable fission fragments that decay to less unstable bodies, then to stable ones. Fission products (FP) are those nuclei whose lifetime is long enough to interest the engineer (minutes, or hours…until stability). A few hundred fission products are retained.

Except for the mass of neutrons and (rare) heavy particles, the total mass of FP is equal to the mass of heavy U and Pu nuclei that have disappeared. There's a whole range of FPs: some absorb, some don't; and of course, some are more or less unstable, with shorter or longer half-lives.

Some FPs are powerful absorbers; two in particular stand out: ^{135}Xe and ^{149}Sm.

Apart from these two, the ten most absorbent FPs alone account for 75% of total absorption."

Figure 12.1 illustrates this statement.

Nuclear reactor operators should be aware of FP poisoning that is an important phenomenon: the loss of core reactivity due to the Xenon effect alone is of the order of 3000 pcm on a PWR, and can rise to over 4000 pcm 7 h after an emergency shutdown.

To understand the Xenon effect, we must first recall the pattern of production and decay of the FPs that generate ^{135}Xe. (See Fig. 12.2).

We specify that for ^{235}U:

6.4% of all fissions produce ^{135}Sb, ^{135}Te or ^{135}I and then ^{135}Xe indirectly.
0.1% of all fissions produce ^{135}Xe directly.

We also specify that the half-life of the β-decay:$^{135}\mathrm{I} \rightarrow {}^{135}\mathrm{Xe}$ is 6.5 h and that the half-life of the β-decay: $\mathrm{Xe}^{135} \rightarrow \mathrm{Cs}^{135}$ is 9 h.

B. Mercier, *Simple Models for Understanding Nuclear Reactor Physics*,
https://doi.org/10.1007/978-3-031-98532-4_12

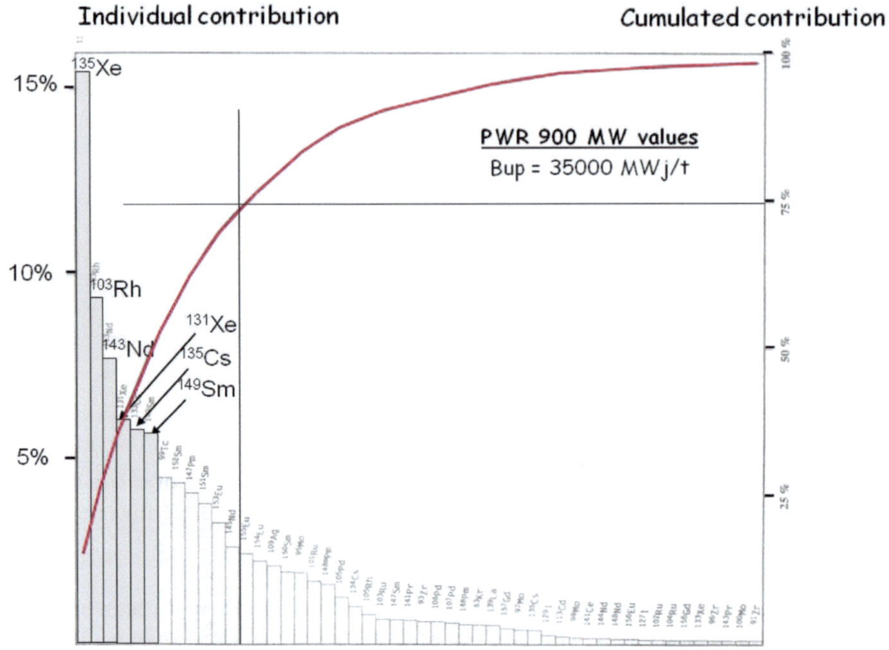

Fig. 12.1 Individual and cumulative absorption of the most absorbing FP

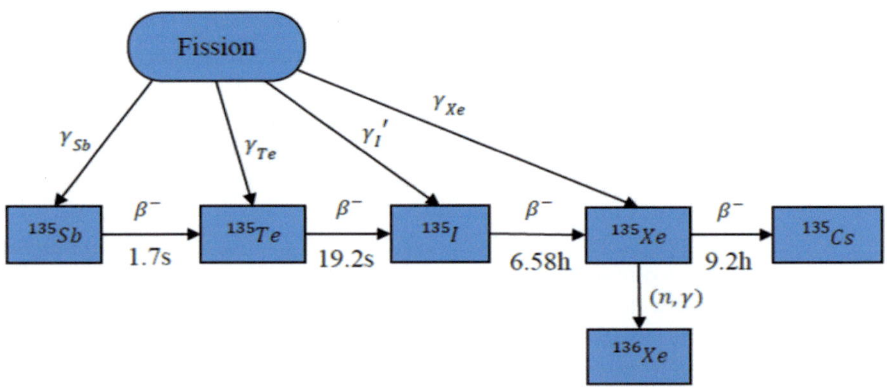

Fig. 12.2 Production and decay of ^{135}Xe

To model the evolution of ^{135}Xe, we generally neglect the very short-lived intermediates ^{135}Sb and ^{135}Te and consider only ^{135}I and ^{135}Xe.

Let I (resp X) denote the number of ^{135}I (resp. ^{135}Xe) atoms per cm^3 in the fuel, the Bateman equations (that we shall study in more detail in Chap. 13) describe their evolution:

$$\frac{\mathrm{d}I}{\mathrm{d}t} + \lambda I = \gamma_I \Sigma_f \Phi \tag{12.1}$$

$$\frac{\mathrm{d}X}{\mathrm{d}t} + (\mu_0 + \sigma\Phi)X = \gamma_X \Sigma_f \Phi + \lambda I \tag{12.2}$$

Both differential equations are coupled, so this is a differential system.

The production term for each of the two species appears on the right-hand side: $\Sigma_f \Phi$ being the number of fissions per second per cm^3 of fuel, $\gamma_I \Sigma_f \Phi$ is then the number of ^{135}I atoms produced per cm^3 per second. Then we have $\gamma_I = 0.064$ and $\gamma_X = 0.001$.

The disappearance term, due to the beta decay of each of these two radioactive elements, is on the left-hand side.

Each time a ^{135}I atom disappears, it produces a ^{135}Xe atom.

Remark 12.1 In (12.1) and (12.2) we can assume either that I (resp.X) is the number of ^{135}I atoms (resp ^{135}Xe) per cm^3 of fuel or per cm^3 of core. Assuming that the neutron flux has the same value in the moderator and in the fuel it will give equivalent results provided we take Σ_f equal to the fuel fission cross section in one case or equal to the core fission cross section in the other case. (ratio $\sim$ 3). ■

When the core is at rated power, the very large capture cross-section of ^{135}Xe ($\sigma = 3 \times 10^6 b$ for thermal neutrons) makes that in steady-state regime:

$$(\mu_0 + \sigma\Phi)X = \gamma_X \Sigma_f \Phi + \lambda I = \gamma_X \Sigma_f \Phi + \gamma_I \Sigma_f \Phi$$

The number X of ^{135}Xe. atoms per cm^3 then stabilizes at a finite value

$$X_\infty = \frac{(\gamma_X + \gamma_I)\Sigma_f\ \Phi}{\mu_0 + \sigma\Phi}$$

Note that X_∞ is increasing with the neutron flux Φ, with an asymptotic value that is $\frac{(\gamma_X+\gamma_I)_f}{\sigma}$.

In the same way the steady state value $I_\infty = \frac{\gamma_I \Sigma_f \Phi}{\lambda}$ for iodin is increasing with the neutron flux Φ. , but there is no asymptotic value as $\Phi \to \infty$.

This explains the Xenon effect: in case of an emergency shutdown, there is a large number of ^{135}I atoms stored in the fuel, it will eventually produce ^{135}Xe and then poison the reactor core.

In fact, when $\Phi = 0$ Eq. (12.2) shows that only $\beta -$ decay can eliminate ^{135}Xe and it is a slower phenomenon than its production by Iodin $\beta -$ decay.

This is illustrated in Fig. 12.3, communicated by Mathilde Moiron that suggests an hydraulic analogy: when the neutron flux $\Phi > 0$ the iodin reservoir fills up. It also empties by $\beta -$ decay, but it fills up to an equilibrium value which depends on Φ.

The Iodin β-decay fills up the Xenon reservoir that also empties due to the sum of two effects (a) β-decay of ^{135}Xe (b) neutron capture when $\Phi > 0$.

Fig. 12.3 Hydraulic analogy

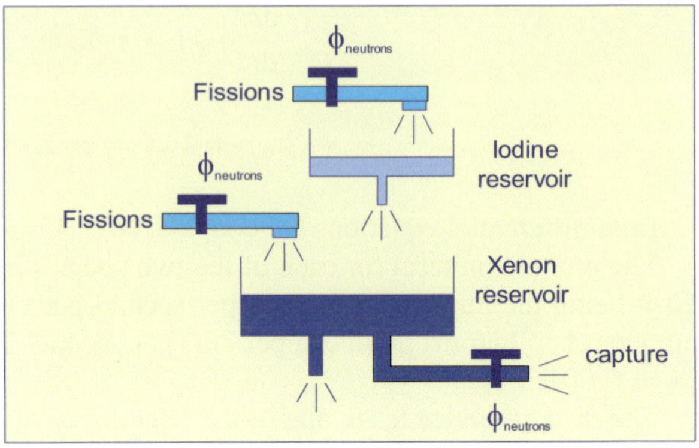

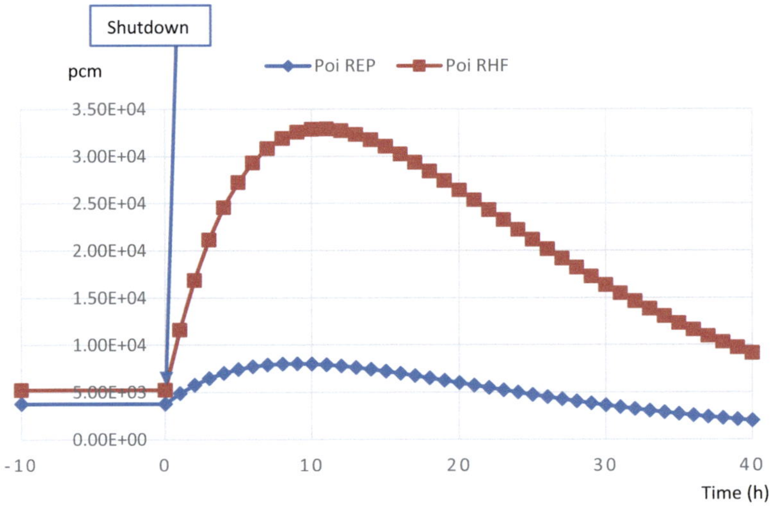

Fig. 12.4 Xenon effect for a PWR compared to a high flux reactor (HF)(The HF curve is drawn with $\Phi = 1.5\ 10^{14}\ \text{n/cm}^2/\text{s}$) shows that the xenon poison effect is much higher on the RHF, so that it is not possible to start again the reactor a few hours after shutdown

When $\Phi = 0$ the iodin reservoir continues to empty at the same flow rate, but the xenon reservoir empties at a smaller flow rate so it fills up.

Another illustration is provided on Fig. 12.4 (also communicated by Mathilde Moiron). We compare the Xenon poisoning after shutdown in a PWR and in a high flux experimental reactor denoted by RHF. Having a highly enriched fuel it can deliver a neutron thermal flux up to $1.5\ 10^{15}\ \text{n/cm}^2/\text{s}$ that is up to 40 times higher than in a PWR.

Evaluation of the Xenon poisoning effect

When there are X atoms per cm^3 of ^{135}Xe in the fuel, its absorbing macroscopic cross section changes from $\Sigma_U \to \Sigma_U + \sigma X$ and then, let Σ_m denote the moderator capture cross section, ω the fuel fraction, $\Sigma_0 = (1-\omega)\Sigma_m + \omega\,\Sigma_U$ the core macroscopic absorbing macroscopic cross section we shall have $\Sigma_0 \to \Sigma_0 + \omega\,\sigma X$.

Given that $\eta = \nu\,\Sigma_f/\Sigma_U$ and $f = \omega\Sigma_U/\Sigma_0$, we see that

$$\eta \to \nu\Sigma_f/(\Sigma_U + \sigma X)$$
$$f \to \omega(\Sigma_U + \sigma X)/(\Sigma_0 + \omega\sigma X)$$

Hence, on the 4 factors of Fermi's formula, two of them change: η and f.

$$k_\infty \to k_\infty \Sigma_0/(\Sigma_0 + \omega\sigma X)$$

The negative reactivity induced by ^{135}Xe is then $10^5 \frac{\omega\,\sigma X}{\Sigma_0}$ pcm.

Exercise 12.1 Consider the differential system

$$\frac{dI}{dt} + \lambda I = f$$
$$\frac{dX}{dt} + \mu X = g + \lambda I$$

For the application to Xenon poisoning we let $\mu = \mu_0 + \sigma\Phi$ with $\mu_0 < \lambda$.

More precisely we take:

$$\lambda = 2.86 \times 10^{-5}\,s^{-1}$$
$$\mu_0 = 2.08 \times 10^{-5}\,s^{-1}$$

$f = \gamma_I \Sigma_f \Phi$ where γ_I is the iodin fission yield

$\Sigma_f\left(cm^{-1}\right)$ the (assumed to be homogeneous) core fission macroscopic cross section.

Φ the neutron flux $\left(n\,cm^{-2}\,s^{-1}\right)$.

$g = \gamma_X \Sigma_f \phi$ where γ_X is the xenon fission yield.

1. Provided f does not depend on time, prove that

$$I(t) = I(0)e^{-\lambda t} + \frac{1 - e^{-\lambda t}}{\lambda} f$$

Deduce that if we start with $I(0) = 0$ we get $I(t) \cong I_\infty = \frac{f}{\lambda}$ after a sufficient time.

Numerical application: take $f = \gamma\Sigma_f$ with $\gamma = 6.4\%$, $\Sigma_f = 0.25\,cm^{-1}$ and $\Phi = 3.66 \times 10^{13} n/cm^2/s$ compute I_∞.

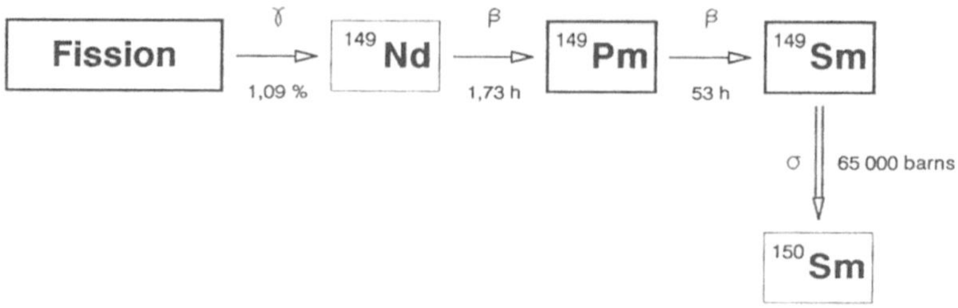

Fig. 12.6 Samarium production process

2. Prove that if we start with $I(0) = 0$ and $X(0) = 0$ then

$$X(t) = \frac{1 - e^{-\mu t}}{\mu}(f + g) - f\frac{e^{-\lambda t} - e^{-\mu t}}{\mu - \lambda}$$

Deduce that $X(t) \to \frac{1}{\mu}(f + g)$ as $t \to \infty$.

Numerical application: take $g = \gamma_X \Sigma_f \Phi$ with $\gamma_X = 0.1\%$, and $\sigma = 3 \times 10^6$ barn; compute g, μ and then X_∞.

3. We now assume $\Phi = 0$ so that $f = g = 0$ (and $\mu = \mu_0$).

Prove that

$$I(t) = I(0)e^{-\lambda t}$$

and

$$X(t) = X(0)e^{-\mu t} + \lambda I(0)\frac{e^{-\lambda t} - e^{-\mu t}}{\mu - \lambda}$$

4. Let $A = \frac{\lambda}{\lambda - \mu} I(0)$ prove that $X(t)$ reaches its maximum when

$$t = \frac{1}{\lambda - \mu}\text{Log}\frac{\lambda A}{\mu X(0) + \mu A}$$

Numerical application: We assume $I(0) = I_\infty$ and $X(0) = X_\infty$ where I_∞ has been calculated at question 1 and X_∞ at question 2, compute t. (*solution p.* 186).

Samarium Poisoning

The production process for Samarium is described in Fig. 12.6.

Like ^{135}Xe, ^{149}Sm is then a poison, but a significant difference is that, unlike ^{135}Xe, ^{149}Sm is stable (no $\beta-$ decay).

Let P (resp. S) denote the number of ^{149}Pm (resp. ^{149}Sm) per cm^3 of fuel, the Bateman equations are then:

$$\frac{dP}{dt} = \gamma \Sigma_f - \lambda P$$
$$\frac{dS}{dt} = \lambda P - \sigma \Phi S$$

where

- γ is the fission yield for ^{149}Nd
- $\lambda = \ln(2)/T$ where $T = 53\,\text{h}$
- $\sigma = 65000b$ is the microscopic capture cross section for ^{149}Sm.

Exercise 12.2 A reactor initially containing neither ^{149}Pm nor ^{149}Sm is started. A constant neutron flux is maintained in the reactor, equal to $2\ 10^{13}\ \text{n/cm}^2/\text{s}$ over time.

Let P (resp. S) denote the number of ^{149}Pm (resp. ^{149}Sm) per cm^3 of fuel, we take for true that the Bateman equations modeling the evolution of P and S are:

$$\frac{dP}{dt} = \gamma \Sigma_f - \lambda P$$
$$\frac{dS}{dt} = \lambda P - \sigma \Phi S$$

where

- Σ_f is the fission macroscopic cross section of the fuel,
- $0 < \gamma < 1$ is the fission yield for ^{149}Nd,
- λ the inverse of the time constant of ^{149}Pm $\beta-$ decay
- σ is the microscopic capture cross section for ^{149}Sm.

We assume the core to be homogeneous: each cm^3 of the core contains ω cm^3 of fuel and $(1 - \omega)$ cm^3 of water. We denote by Σ_a the absorption macroscopic cross section of the core, taking into account the boron in the moderator.

1. Explain why, when $t \to \infty$ the Samarium brings a negative reactivity equal to $\frac{\omega\, \gamma\, \Sigma_f}{\Sigma_a} 10^5$ pcm.
2. Numerical application: with 1000 ppm boron in the moderator we take $\Sigma_a = 0.276\text{cm}^{-1}$ $\Sigma_f = 0.190\,\text{cm}^{-1}$ and $\gamma = 0.0109$. Compute the negative reactivity due to Samarium.
3. Assuming $\sigma = 65{,}000\ b$ compute S in steady state regime.
4. The half-life of ^{149}Pm is assumed to be 53 h; compute P in steady state regime.
5. What happens in case of an emergency shutdown? And when the reactor restarts? (*solution p.* 188).

Unlike Xenon, which provides a negative reactivity that disappears after a sufficiently long pause, the poisoning induced by the presence of Samarium is long-lasting.

Exercise 12.3 Rhodium $^{103}_{45}$Rh is a fission product whose fission yield is $\gamma = 0.06$ and the microscopic capture cross section is $\sigma = 500b$.

Unlike $^{135}_{54}$Xe, this fission product is stable (no radioactive decay).

1. Let R denote the number of $^{103}_{45}$Rh atoms per fuel cm^3, and Σ_f^U the macroscopic fission cross section of the fuel, explain how we obtain the differential equation:

$$\frac{\mathrm{d}R}{\mathrm{d}t} = \gamma \Sigma_f^U \Phi - \sigma \Phi R \tag{12.3}$$

2. We consider a steady state regime where the neutron flux Φ is independent of time. Assuming $R(0) = 0$, find the solution $R(t)$ of the differential Eq. (12.3). Deduce that if $t \to \infty$, $R(t) \to \gamma\ \Sigma_f^U/\sigma$.
3. In steady-state regime, calculate (in pcm) the decrease in reactivity (i.e., poisoning) caused by this fission product. (take $\Sigma_f^U = 0.527\text{cm}^{-1}, \omega = 0.33$ and $\Sigma_0 = 0.34\text{cm}^{-1}$ where Σ_0 is the core macroscopic absorption cross section).■ (*solution p.* 189).

Chapter 13
Fuel Evolution and Management

The energy contained in 1 g of fissile material being about equivalent to the energy contained in 1 ton of oil, refueling a Nuclear Power Plant occurs about every year. But it takes about one month, so such refueling phases should be carefully planned. This is why we shall dedicate this chapter to fissile material consumption and fuel evolution that leads to reactivity decay of fuel assemblies. Ideally, refueling should occur when the reactivity of the core is not sufficient for the core to diverge. (The elapse time between two refuelings is called a cycle).

The reactor core is the site of fission and capture reactions, but $(n, 2n)$ reactions can also occur. In addition, most of the nuclei present in the core are radioactive, and can therefore generate other elements through radioactive decay.

Figure 13.1 is taken from [10]:

Columns show atomic numbers from Uranium ($Z = 92$) to Curium ($Z = 96$). The rows show the atomic mass from $A = 235$ to $A = 244$.

Neutron capture moves from one row to the next without changing columns. A β decay moves from one column to the next without changing rows. Electron capture (EC) from one column to the next.

On Fig. 13.1 we can clearly see the successive reactions whereby the capture of a neutron by ^{238}U will lead (albeit not instantaneously) to the creation of ^{239}Pu. Plutonium is a metal that does not occur naturally and can only be produced in a reactor.

As the fuel in the core changes, so does the core's reactivity.

Figure 13.2 is reproduced from [2].

The x-axis shows the burn-up rate. The burn-up is evaluated in GWd/t or, equivalently, in MWd/kg. This characterizes the energy produced per unit mass of heavy metal (Uranium in the case of UO_2 fuel).

The y-axis shows the mass ratio of heavy nuclei in logarithmic scale.

As an example, given that a PWR900 contains 79,600 kg of heavy metal, the daily burn-up is equal to 2700/79600 = 0.0339 MWd/kg. (Reminder: $1\,\text{MWd} = 10^6 \times 24 \times 60 \times 60\,\text{J} = 86.4\,\text{GJ}$).

B. Mercier, *Simple Models for Understanding Nuclear Reactor Physics*,
https://doi.org/10.1007/978-3-031-98532-4_13

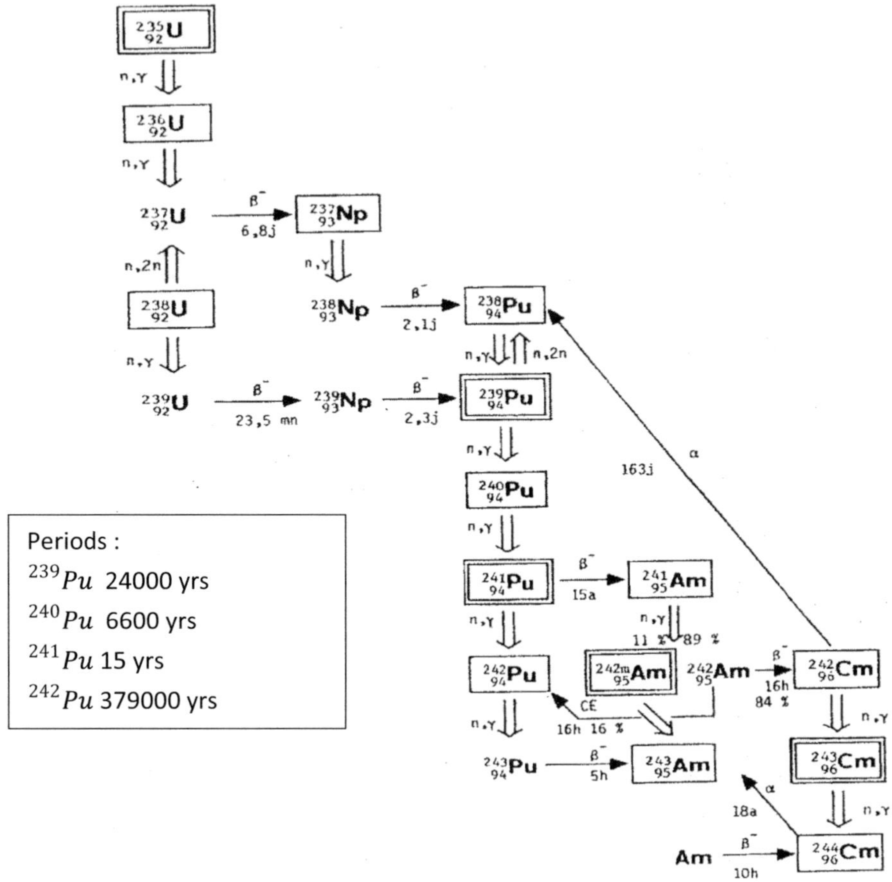

Fig. 13.1 Uranium evolution chain

One objective for EDF in France is to reach a burn-up of 52 GWd/t, hence 1 fuel assembly should remain 1534 full power equivalent days (FPED) in the reactor (result of the ratio 52/0.0339).

We can see on Fig. 13.2 that ^{239}Pu inventory rises relatively rapidly until it reaches an asymptote, at which point every new plutonium nucleus produced is consumed. In contrast, the ^{235}U inventory decreases until the two curves cross.

Figure. 13.3, taken from [10], shows that, for a PWR, the k_∞ of an assembly decreases linearly

This figure, which is a simplification of reality, does not take into account poisoning by fission products, which is of the order of 4000 pcm from the 3rd FPED (full power equivalent day).

Note that the decrease is about 30,000 pcm for a 30 GWd/t burn-up.

Since we reason at the level of an assembly, it's the assembly's k_∞ that matters.

We'll sometimes get a $k_\infty < 1$ in parts of the core that have seen 3 or 4 cycles....

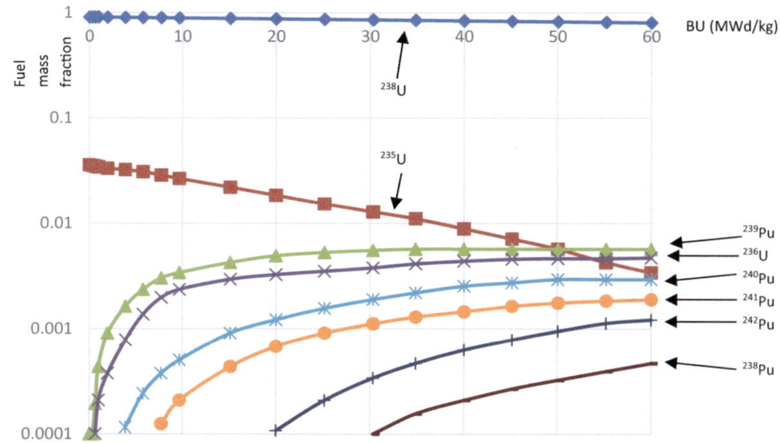

Fig. 13.2 Heavy nuclei evolution w.r.t. burn-up

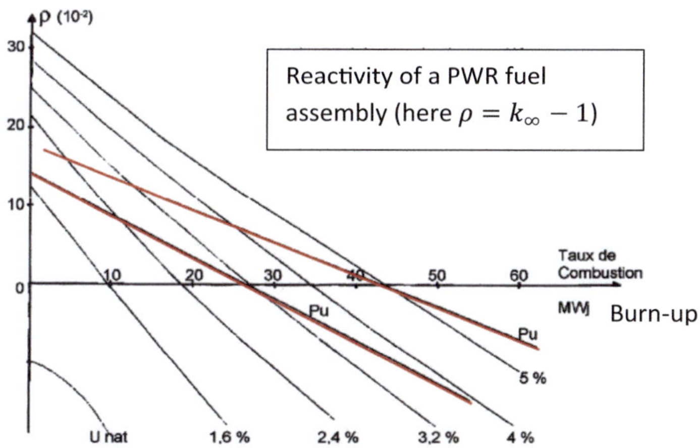

Fig. 13.3 Reactivity evolution as a function of burn-up

In any case, over the course of the reactor cycle, the lower reactivity of the assemblies will have to be compensated for by reducing the boron concentration: this is 1500ppm at BOC (Beginning of Cycle) 0 ppm at EOC (end of cycle). If we can no longer diverge with pure water, the cycle will come to an end. To plan NPP shutdowns properly, you need to avoid unpleasant surprises.

Terminology: during the reactor cycle there is *consumption* (of fissile nuclei or consumable poisons), *conversion* (of fertile nuclei), *poisoning* (by absorbing nuclei, e.g., FP).

BOC means beginning of cycle, and EOC = End of Cycle.

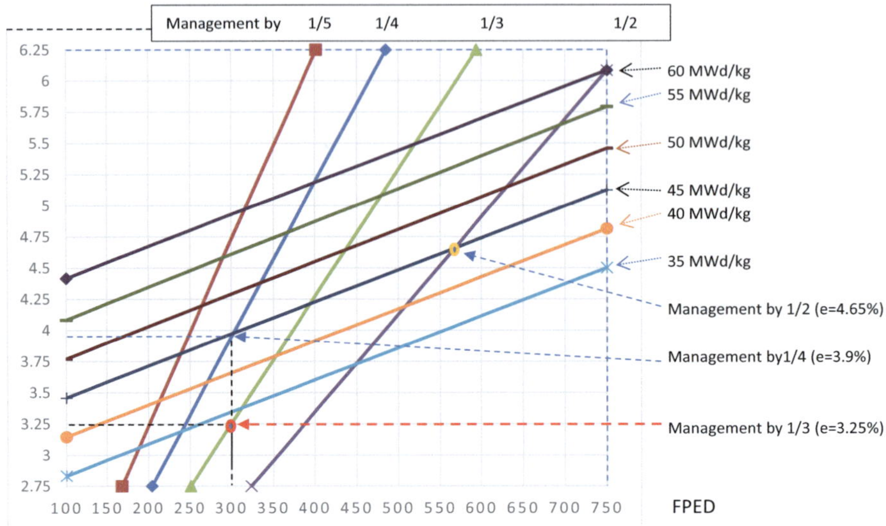

Fig. 13.4 Management strategies for fuel assemblies [10]

For the NPP operator an issue is to find a good compromise between the fuel enrichment, the fuel burn-up and the length of a cycle. This is represented on Fig. 13.4 extracted from [10].

On Fig. 13.4 the cycle length (in FPED) appears on the *x*-axis. The fuel enrichment is on the *y*-axis. Parallel oblique lines show the burn-up in MWd/kg. On a 900 MWe there are 157 fuel assemblies.

The core can be managed by half: at the end of each cycle, 76 fuel assemblies that have been irradiated for 2 cycles in the core are replaced by new ones. To reach a burn-up of 45 GWd/t it shows that both an enrichment of 4.65% and a cycle length of 570 FPED are needed.

The core can also be managed by quarter: at the end of each cycle, 40 fuel assemblies that have been irradiated for 4 cycles in the core are replaced by new ones.

Figure 13.4 shows that with such a management, to reach the same burn-up, an enrichment of 3.9% only is needed. The cycle length will be equal to 300 FPED which is compatible with a yearly cycle.

A management by third has been used for a long time by EDF: with a cycle length of 300 FPED and an enrichment of 3.25% it is possible to reach a burn-up of 34 GWd/t.

Increasing the fuel burn-up is a good target. However Figs. 13.1 and 13.2 show that there are some drawbacks:

- the initial enrichment must be increased

Fig. 13.5 Fuel management strategy by quarter (from [2])

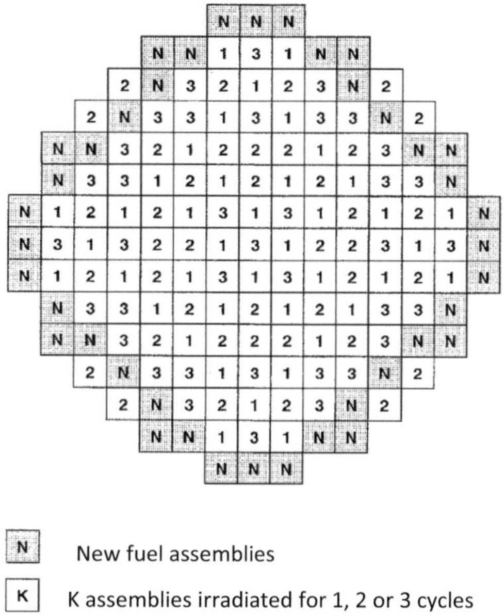

- the irradiated fuel will contain more transuranics (like ^{242}Pu or ^{238}Pu) and minor actinides (like ^{244}Cm). This will induce some consequences at the used fuel reprocessing level (Fig. 13.5).

Evolution modeling

The Batemann equations are used to describe the fuel evolution; they all have the same structure: following Figure 13.6 we assume that A $\rightarrow$ C and C $\rightarrow$ E via neutron capture and that $B \rightarrow C$ and $C \rightarrow D$ via $\beta-$ decay. The Bateman equation describing the evolution of nucleus C can be written

$$\frac{d}{dt}N_C = \lambda_B N_B + \sigma_{cA} N_A \Phi - \lambda_C N_C - \sigma_{cC} N_C \Phi \qquad (13.1)$$

where Φ is the local neutron flux. For all nuclei appearing on Fig. 13.1, a Batemann equation has to be written, and all of them are coupled.

Furthermore the Batemann equations should be coupled with the Boltzmann equation used to compute the neutron flux. Solving the Batemann equations will give the fuel composition and then the macroscopic cross sections which are needed to solve the Boltzmann equation. Conversely, solving the Boltzmann equation will give the neutron flux.

Remark 13.1 Bateman's equation (13.1) also applies to fission products and their progeny. However, for a fission product, as we saw in Exercises 12.1 and 12.2, a source term of the form $\gamma \Sigma_f \Phi$ should be added.

Fig. 13.6 Scheme used to introduce the Bateman equations

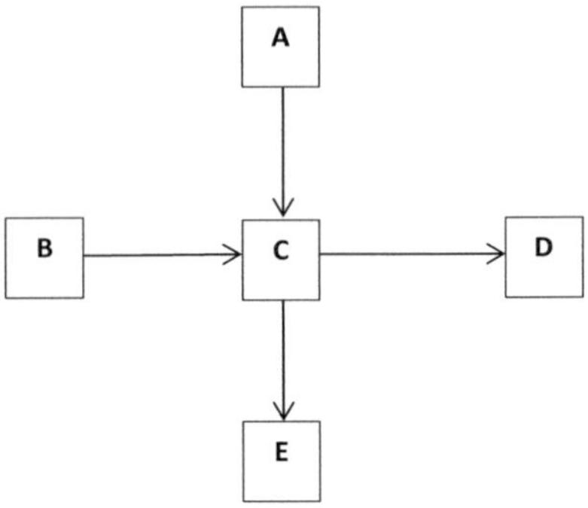

Exercise 13.1

1. We denote by $M_9(t)$ (resp. $N_9(t)$) the number of ^{239}Np (resp ^{239}Pu) at/cm^3 in the fuel located in the core of a PWR. By what reasoning can we establish the system of equations:

$$\frac{d}{dt} M_9 + \lambda M_9 = \Sigma_{c8} \Phi \tag{13.2}$$

$$\frac{d}{dt} N_9 + \mu N_9 = \lambda M_9? \tag{13.3}$$

   How to choose λ et μ?

2. We assume that N_8 Φ are constant (this is an approximation for N_8) and that $M_9(0) = N_9(0) = 0$. Show that $M_9(t) \to M_9^\infty$ and $N_9(t) \to N_9^\infty$ while $t \to \infty$. Compute M_9^∞ and N_9^∞ with respect to the data provided in the exercise.
3. Deduce that there are as many ^{238}U nuclei that produce ^{239}Pu as ^{239}Pu nuclei that disappear by capture or fission.
4. Prove that

$$\frac{d}{dt} N_9 + \mu N_9 = \Sigma_{c8} \Phi \tag{13.4}$$

   Is a good approximation to (13.2) and (13.3).

   Numerical data: $\sigma_{a9} = 1000b$, $\Phi = 3.66 \times 10^{13}$ n/cm^2/s, $N_8 = 0.02445\ 10^{24}$ at/cm^3,
   $\sigma_{c8} = 2.7b$. Period for ^{239}Np $\beta-$decay $T = 55.2$ h. (*solution* p. 190).

Remark 13.2 For simplicity we have neglected the ^{239}U intermediate. The reason is that its period is 140 times shorter than that of ^{239}Np. Since Exercise 13.1 shows that for a reasonable number of FPED we can neglect the ^{239}Np intermediate, we can even more neglect ^{239}U intermediate.

Exercise 13.2 We shall take for true that

1. fission of heavy nuclei produces 82 GJ/g, be it for ^{235}U, ^{238}U, ^{239}Pu or ^{241}Pu.
2. on a PWR900 (whose power is 2.7 GWth, fast fissions consume 3.4 g of ^{238}U per FPED.

1. The conversion factor of a PWR is $c = 0.6$: each time 1 g of ^{235}U is consumed, 0.6 g of Pu is produced. However, only a fraction $p_1 = 0.85$ of the ^{235}U atoms consumed are fissioned (the others giving ^{236}U after neutron capture), and only a fraction $p_3 = 0.73$ of the plutonium atoms produced (by decay of the ^{239}U produced by capture) are fissioned (either as ^{239}Pu or as ^{241}Pu).

   Deduce the mass δm of ^{235}U consumed each FPED on a PWR 900.

   (We neglect the ^{239}U–> ^{239}Np and ^{239}Np $\rightarrow$ ^{239}Pu $\beta -$. decay (i-e., we assume that neutron capture by ^{238}U directly produces ^{239}Pu).
2. The initial Uranium enrichment is assumed to be $e = 4\%$. We denote by M the Uranium mass initially contained in the core. We assume $M = 79.6$ t which corresponds to a PWR900.

   We start from a new core. We denote by m the ^{235}U mass initially available and by m_0 the ^{235}U below which the reactor will not diverge.

   We shall take for true that $m_0 = 1468$ kg.

   Compute the number x of FPED which can be produced with this new core.

   What will be the residual ^{235}U enrichment?

   Evaluate the fuel Burn Up (in GWd/t).
3. We adopt a core management by half. Let x denote the cycle duration (in FPED). We then start with a new half core and another one having already been irradiated during x FPED.

   What is the greatest value for x?

   What will be the residual ^{235}U enrichment, at the end of the second cycle for the most used half core?

   Evaluate its Burn Up?
4. The core is now managed in n equal sub-parts. We start the cycle with fuel that has undergone $(i - 1)$ cycles in the reactor at BOC in subpart $n°i$. At EOC (end of cycle), the fuel in sub-part $n°i$ has already undergone i reactor cycles, $i = 1, \ldots, n$

   With this type of management, what could be the duration x of the cycle in FPED?

   What is the residual enrichment of the most depleted part?
5. Let E_n denote the energy provided by a fuel assembly by using a core management with n sub-parts prove that we have

$$E_n = 2n/(n + 1) \cdot E_1$$

Deduce the Burn Up that can be obtained with $n = 3$ then $n = 4$. (*solution* p. 191).

The conclusion of this exercise is that for a given enrichment the higher n the higher the Burn Up. The counterpart is that the higher n, the shorter the cycle duration. So the operator should find an adequate compromise.

Exercise 13.3 [10] The following table gives the initial and final composition of a PWR fuel (in kg per t of heavy metal).

	Initial	Final
U235	45	9
U236	0	5.49
U238	955	910
Pu239	0	5
Pu 240	0	2.4
Pu 241	0	2.2
Pu 242	0	0.94

We neglect absorption by ^{236}U, and ^{242}Pu.
We neglect the ^{241}Pu $\beta -$ decay.
For one ton of initial heavy metal,

1. We assume that 6 kg of ^{238}U have been fissioned. What is the mass of ^{239}Pu produced by capture on ^{238}U?
2. Knowing that the capture/fission ratio is 0.33 for ^{239}Pu, what is the mass of fissioned ^{239}Pu?
3. What is the mass of ^{240}Pu produced? ^{241}Pu produced
4. What is the mass of fissioned ^{241}Pu?
5. what is the mass of fissioned ^{235}U?
6. what is the mass of fissioned nuclei?
7. what is the burn-up of the used fuel ? (*solution* p. 193).

Exercise 13.4 Like [9, p. 382], we consider 4 neutron groups:

1. Fast > 0.9 MeV
2. Fast between 2 keV and 0.9 MeV
3. Epithermal between 6 eV and 2 keV
4. Thermal < 6 eV

We shall take for true the following average values for the microscopic cross sections of ^{238}U and ^{235}U:

	Gr 1	Gr 2	Gr 3	Gr 4	Units
σ_f^8	1	0	0	0	b
σ_c^8	0	0	4.4	2.7	b
σ_f^5	0	1.85	27.06	271	b
σ_c^5	0	0.5	25.2	96	b

Given that $N_8 = 0.02364\,\text{at/b/cm}$ and $N_5 = 0.000875\,\text{at/b/cm}$ we deduce the following macroscopic cross sections.

On the 5th line of the table below, we indicate the neutron flux integrated in each group. Fissions and captures for ^{238}U and ^{235}U that are given on lines 6–9, are easily deduced from lines 1–5.

	Gr 1	Gr 2	Gr 3	Gr 4	Units
Σ_f^8	0.02364	0	0	0	cm^{-1}
Σ_c^8	0	0	0.13	0.06383	cm^{-1}
Σ_f^5	0	0.00162	0.0236	0.3849	cm^{-1}
Σ_c^5	0	0.000437	0.022	0.0525	cm^{-1}
Φ	2.8×10^{13}	11.4×10^{13}	6.4×10^{13}	3.66×10^{13}	$n/cm^2/s$
FissionsU8	$0.662 10^{12}$	0	0	0	$f/cm^3/s$
Fissions U5	0	$0.184 10^{12}$	$1.51 10^{12}$	$14.1 10^{12}$	$f/cm^3/s$
Captures U8	0	0	$8.32 10^{12}$	$2.34 10^{12}$	$c/cm^3/s$
Captures U5	0	$4.99 10^{10}$	$1.41 10^{12}$	$1.92 10^{12}$	$c/cm^3/s$

1. Using the above table compute the conversion factor C.
2. Calculate factors ε, p and η of Fermi's formula.
3. Given that for light water at the operating PWR temperature we have $\Sigma_c^H = 0.01\,\text{cm}^{-1}$ and that the moderation ratio $V_m/V_U = 2$ compute the 4th factor f. Deduce the value of k_∞. (*sol.* p. 193).

Thorium evolution chain.

The main isotope available in natural thorium is ^{232}Th. However it is not fissile by thermal neutrons.

On Fig. 13.7 we see that it can lead to

- ^{231}Th via $(n, 2n)$ reaction.
- ^{233}Th after neutron capture.

We can see that $^{233}\text{Th} \rightarrow ^{233}$ Pa after $\beta-$decay (period $T = 23$ mn) and then ^{233}U ($T = 27.4$ days).

Since ^{233}U is a fissile material, ^{232}Th is a *fertile* material (like ^{238}U) and that's why it is very useful.

Remarks 13.3

- Thorium is 3 times more abundant on earth than uranium.
- ^{233}U is fissile by thermal neutrons so it could be used in PWRs.
- ^{233}U does not exist in nature. It has to be produced in a nuclear reactor using another fissile material (^{235}U or ^{239}Pu) preferably in Fast Nuclear Reactors.

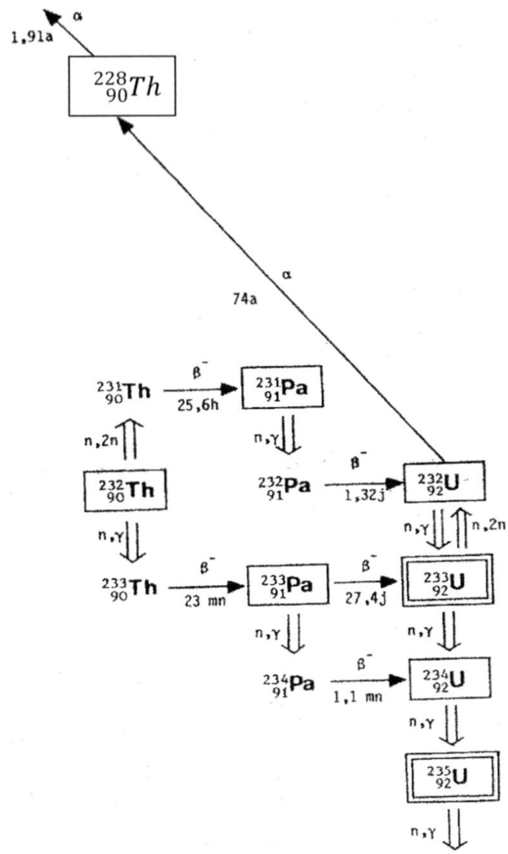

Fig. 13.7 Thorium evolution chain. From [7]

Table 13.1 Conversion factor for different types of reactors

Type	*C*
NUGG	0.8
CANDU	0.9
PWR	0.6
FNR	0.8
FNR + AB	1
FNR + AB + RB	1.2

Conversion factor

It is the ratio C between the number of fissile nuclei produced (per s) divided by the number of fissile nuclei consumed (per s). See Table 13.1 extracted from [10], where NUGG = Natural Uranium Graphite Gas, AB means Axial (fertile) Blanket and RB Radial (fertile) blanket.

Exercise 13.5

1. Write Bateman's equation for the ^{233}Pa evolution
2. Assume that the microscopic capture cross section of ^{233}Pa is equal to $43b$, for which neutron flux Φ will the ^{233}U production be equal to the ^{234}Pa production? (*solution* p. 194)

Chapter 14
Criticality Risk Control

The reactor core is a device where criticality is expected.

On the contrary, outside, and in particular in the fuel cycle facilities, criticality should be avoided by all means.

We shall see that the operator should control

1. The mass of fissile material.
2. Its geometry: as we shall see, slender geometries foster neutron leakage. Sphere is the most penalizing geometry.
3. Its moderation: with a relatively small mass of fissile material criticality is possible with a sufficient amount of moderator.

Critical Mass

We have introduced one and two groups diffusion models in which the fundamental eigenvalue λ_0 is used to evaluate the non-leakage probability [see (10.5) or (10.6)].

We shall see how it depends on geometry.

We first consider the sphere since we said it's the most penalizing geometry.

As we have seen in Chap. 9 to calculate λ_0, we must solve the eigenvalue problem

$$-\Delta\Phi_0 = \lambda_0\Phi_0 \quad \text{for}\, x \in V \tag{14.1}$$

$$\Phi_0 \geq 0 \quad \text{for}\, x \in V \tag{14.2}$$

$$\Phi_0 = 0 \quad \text{for}\, x \in V \tag{14.3}$$

In spherical coordinates $\{r, \theta, \varphi\}$ we have

$$\Delta = \frac{\partial^2}{\partial r^2} + \frac{2}{r}\frac{\partial}{\partial r} + \frac{1}{r^2}\left(\frac{1}{\sin\theta}\frac{\partial}{\partial\theta}(\sin\theta)\frac{\partial}{\partial\theta} + \frac{1}{\sin^2\theta}\frac{\partial^2}{\partial\varphi^2}\right)$$

B. Mercier, *Simple Models for Understanding Nuclear Reactor Physics*,
https://doi.org/10.1007/978-3-031-98532-4_14

an expression that simplifies considerably when applied to a function $\Phi_0 = \Phi_0(r)$ that depends neither on θ nor on φ.

When V is a sphere of radius R, let

$$\Phi_0(r) = \frac{1}{r}\sin\left(\frac{\pi r}{R}\right)$$

we check that

$$\begin{aligned}
\Phi_0(r) &\geq 0 \quad \text{for } 0 \leq r \leq R \\
\Phi_0(0) &= \frac{\pi}{R} \\
\Phi_0(R) &= 0 \\
\frac{\partial}{\partial r}\Phi_0 &= \frac{\pi}{R}\frac{1}{r}\cos\left(\frac{\pi r}{R}\right) - \frac{1}{r^2}\sin\left(\frac{\pi r}{R}\right) \\
\frac{\partial^2}{\partial r^2}\Phi_0 &= -\left(\frac{\pi}{R}\right)^2\frac{1}{r}\sin\left(\frac{\pi r}{R}\right) - \frac{2}{r^2}\frac{\pi}{R}\cos\left(\frac{\pi r}{R}\right) + \frac{2}{r^3}\sin\left(\frac{\pi r}{R}\right) \\
\Delta\Phi_0 &= \frac{\partial^2}{\partial r^2}\Phi_0 + \frac{2}{r}\frac{\partial}{\partial r}\Phi_0 = -\left(\frac{\pi}{R}\right)^2\frac{1}{r}\sin\left(\frac{\pi r}{R}\right) = -\left(\frac{\pi}{R}\right)^2\Phi_0(r)
\end{aligned}$$

We deduce that $\lambda_0 = \left(\frac{\pi}{R}\right)^2$.

Of course formulae (10.6) and (10.7) show that the non-leakage probability also depend on the migration area M^2, but the latter depends on the material only.

For a given material, having λ_0 large reduces the non-leakage probability and then reduces the criticality risk.

Slab/Sphere Comparison

For a slab with thickness $2a$ along the z-axis but infinite along the x and y-axis we have seen in (9.20) that the fundamental eigenvalue is $\lambda_0 = (\pi/2a)^2$; it is the same as for a sphere of radius $R = 2a$; however, unlike the sphere the slab contains an infinite mass.

Even for a parallelepiped, formula (9.20) shows that for the same value of λ_0 a parallelepiped (even a cube) will induce a larger critical mass.

If there is a significant mass of fissile material, it should not be placed in the form of a sphere.

Geometric control involves placing the fissile material in slender containers.

Other Possible Measures

The one group diffusion model tells us that to eliminate the criticality risk we should have

$$k_\infty < 1 + \lambda_0 M^2$$

that will induce $k_{\text{eff}} < 1$.

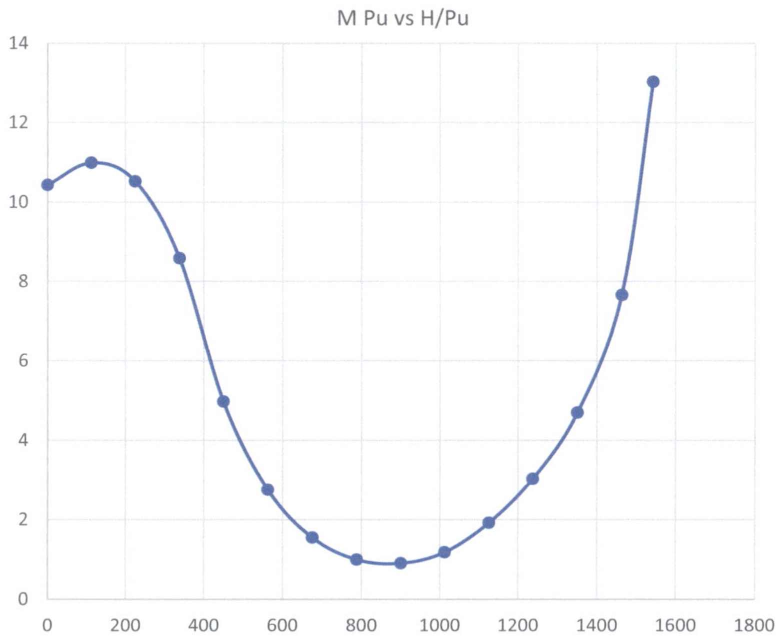

Fig. 14.1 Pu critical mass w.r.t. dilution ratio H/Pu, from [12]

If we cannot change λ_0 we may try to reduce k_∞.

When the fissile material contains some (assumed homogeneously distributed) moderator, we have

$$k_\infty = \varepsilon \,.\, p \,.\, f \,.\, \eta$$

To reduce k_∞.we can reduce p by decreasing the moderation ratio, or, reduce f by increasing the moderator in case it is an absorbing one. We could also decrease f by introducing neutronic poison in the moderator. We could also decrease η by decreasing the enrichment.

This is explained in more detail in an IRSN movie [11].

On Fig. 14.1, extracted from [12] the fissile material is ^{239}Pu, in the form of PuO_2.

The x-axis shows the dilution ratio N_m/N_0 = H/Pu (using the notations from Chap. 7).

We can see that the "critical" mass that leads to $k_{\text{eff}} = 1$ in the most penalizing geometry, i.e., the spherical geometry, depends strongly on the dilution ratio.

If the dilution ratio is too high, the capture cross-section of hydrogen wins out.

As long as the dilution ratio is below 900, the moderating effect dominates.

The optimal dilution is obtained for H/Pu $\sim$ 900 and gives a critical mass equal to 0.9 kg.

For dry powder the critical mass is equal to 10.4 kg.

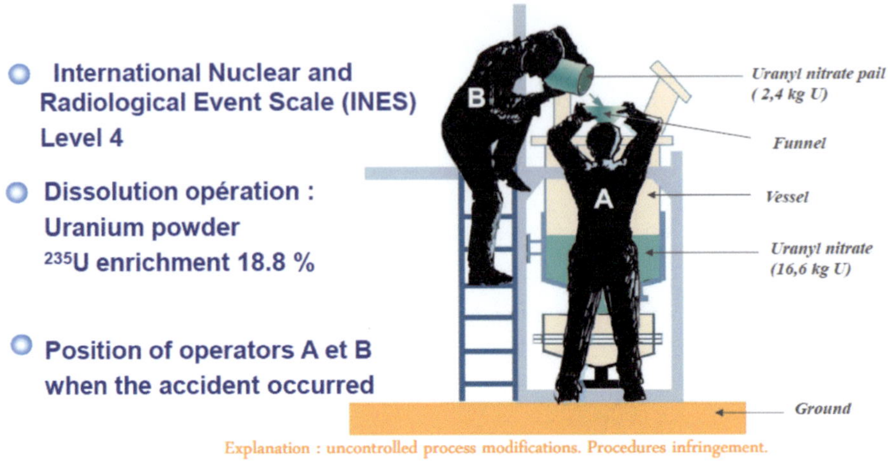

Fig. 14.2 The Tokai-Mura accident

The reader will note that for dilution ratios close to zero, the curve increases slightly, up to a critical mass of 11 kg for H/Pu $\sim$ 100.

This is because, when there is a little moderator, neutrons are slowed down enough to arrive in the ^{239}Pu capture peak zone (numerous peaks at 1000 *b*) but not slowed down enough to arrive in the thermal zone, where the fission cross section is between 500 and 1000 *b*.

Cf. https://www.oecd-nea.org/janisweb/index.html.

In summary, with PuO_2 powder, we obtain, in case of spherical geometry:

- $k_{eff} = 1$ with 10.4 kg for dry powder
- $k_{eff} = 1$ with 0.9 kg for powder in aqueous solution to the optimum of moderation
- $k_{eff} = 1$ with 0.51 kg for powder in aqueous solution with an additional 20 cm-thick water reflector surrounding the sphere.

Precautions must therefore be taken when handling a mass greater than 500 g of PuO_2.

Possible precautionary measures are as follows:

- Control of fissile material mass
- Insert neutron poisons B, Cd, Gd, Hf
- Promote neutron leakage (geometry control)
- Avoid reflectors (concrete walls, etc.)
- Separate containers with neutron shields.

The Tokai-Mura Accident (09/30/1999)

Figure 14.2 is extracted from [12].

The consequences of this accident were severe (2 deaths, 133 people irradiated).

There is now consensus on the causes of the accident:

a. No display of mass limitation instructions.
b. No criticality expert on site.
c. Operators not made aware of criticality risk.

There has been both uncontrolled process modifications and procedure infringement.

(a) uncontrolled process modifications:

- the initial dissolver had a (slender) safe geometry
- somebody changed it to a large precipitator/decanter
- somebody introduced a water cooling of the precipitator/decanter.

(b) procedure infringement

- there was a UO_2 mass limit of 2.4 kg, the operators went up to 16 kg.

Exercise 14.1 We consider a sphere $\{|r| \leq R\}$ and we assume that it contains an homogeneous medium such that each cm^3 contains $\omega\,\text{cm}^3$ of uranium oxyde UO_2 and $(1-\omega)\,\text{cm}^3$ of light water H_2O.

To evaluate the critical mass of this homogeneous medium, we first consider the one group model:

$$-D\left(\Phi'' + \frac{2}{r}\Phi'\right) + \Sigma_0\Phi = \nu_e\Sigma_f\Phi \quad \text{for } |r| \leq R. \tag{14.4}$$

$$\Phi'(0) = 0 \tag{14.5}$$

$$\Phi(R) = 0 \tag{14.6}$$

where $\nu_e = \varepsilon \cdot p \cdot \nu$

1. How should we choose Σ_0 and Σ_f in Eq. (14.4) so that it correctly models the physical situation considered here?
2. Numerical data: for light water we take $\Sigma_c^H = 0.0215\,\text{cm}^{-1}$ and $\Sigma_s^H = 1.34\,\text{cm}^{-1}$; for 3.7% enriched UO_2 we take $\Sigma_a^U = 0.681\,\text{cm}^{-1}$ and $\Sigma_f^U = 0.527\,\text{cm}^{-1}$.

   We choose $\omega = 0.2$, compute Σ_0 et Σ_f.
3. Prove that, in our homogeneous medium, there are about 11 hydrogen atoms for 1 uranium atom. The microscopic scattering cross section of hydrogen being $\sigma_{sH} = 20\,b$. Compute the dilution cross section σ_d then the resonance escape probability p by using some results given in Chap. 7.

   Then assuming $\varepsilon = 1.05$ compute $\nu_e = \nu\,.\,\varepsilon\,.\,p$ and k_∞.
4. We take $M^2 = 50\,\text{cm}^2$. Find the critical radius R. Deduce the critical mass of Uranium m_1 given by our one group model.

5. We now consider the two groups diffusion model studied in Chap. 10. We remind the critical Eq. (10.5):

$$\frac{k_\infty}{\left(1 + M_1^2\lambda_0\right)\left(1 + M_2^2\lambda_0\right)} = 1$$

We choose $M_2^2 = 5\,\text{cm}^2$ for the thermal group and $M_1^2 = 45\,\text{cm}^2$ for the fast group. Being given the value of k_∞ compute λ_0. Deduce the critical radius, then the critical mass m_2 given by our two groups model. Compare with the result obtained at question 4 (*solution*, p.195). ■

Exercise 14.2 Let V denote the sphere of radius R, centered at the origin. Its volume is then equal to $\frac{4}{3}\pi R^3$.

Let Γ denote its boundary: its surface is then equal to $4\pi R^2$.

We shall take for true that the smallest $\lambda > 0$ such that there exists a nonzero function $\Phi_0 = \Phi_0(x)$ verifiying

$$\begin{aligned} -\Delta\Phi_0 &= \lambda\Phi_0 \quad \text{for}\, x \in V \\ \Phi_0 &= 0 \quad \text{for}\, x \in \Gamma \\ \Phi_0(x) &\geq 0 \quad \text{for}\, x \in V \end{aligned}$$

is $\lambda = \lambda_0 = (\pi/R)^2$.

We assume that V is filled with PuO_2 powder of density 11.5 g/cm^3, plutonium being made up solely of ^{239}Pu whose fission, absorption and elastic collision microscopic cross sections are assumed to be, respectively:

$$\begin{aligned} \sigma_f^9 &= 1.9\,b \\ \sigma_a^9 &= 1.96\,b \\ \sigma_s^9 &= 10\,b \end{aligned}$$

Note that these values are valid for fast neutrons for which $\nu = 2.91$.

1. Compute the macroscopic cross sections Σ_f (fission), Σ_a (absorption) Σ_s (elastic collision) for such a PuO_2 powder, given that for oxygen $\sigma_f^O = 0$, $\sigma_a^O \cong 0$ and $\sigma_s^O = 4\,b$.
2. Compute the neutron mean free path in V.

   In case of an event during its random walk in V, calculate the probability that this event is (a) an elastic collision on an oxygen atom? (b) an elastic collision on a ^{239}Pu atom? (c) an absorption by the ^{239}Pu?
3. Why do you think thermal neutrons can be neglected?
4. To find out whether the sphere is critical, we know that we just have to find a positive solution Φ to the problem.

$$
\begin{aligned}
& -D\Delta\Phi + \Sigma_a\Phi = \nu\,\Sigma_f\Phi \quad \text{for}\, x \in V \\
& \Phi = 0 \quad \text{for}\, x \in \Gamma
\end{aligned}
$$

Prove that we must have

$$(*) \quad k_\infty = 1 + \lambda_0 M^2$$

with $M^2 = D/\Sigma_a$ and $k_\infty = \nu\Sigma_f/\Sigma_a$.

5. Compute k_∞ (with at least 3 digits).
6. Compute the diffusion coefficient D, by using a formula given in this textbook, then the migration area M^2.
7. Deduce the value of the critical radius, then the critical mass.
8. Could this result be improved by using the extrapolation length, and how? (*solution*, p.196). ■

Chapter 15
Temperature Effects

In this chapter we show how important are negative reactive effects like the Doppler effect and the moderator effect: they make the reactor stable.

In a PWR there are two significant effects.

- The Doppler effect: as fuel temperature rises, the ^{238}U resonances (or traps) widen. The neutron captures during the slowing down phase increase. The p-factor decreases. Then k_∞ decreases which produces a negative variation of reactivity. The physical basis for the Doppler effect is explained in [2] and in Chap. 7.
- The fact that a PWR is *undermoderated*: i.e., the moderator to fuel ratio is chosen smaller than the optimal value. When temperature increases water expands so that the number of hydrogen atoms per cm^3 decreases and neutrons have fewer elastic collisions allowing them to be thermalized: p decreases but f increases. For an undermoderated reactor the product $p.f$ decreases, as we have seen on Fig. 7.2.

Let T_f denote the average fuel temperature, T_m the average moderator temperature and ρ reactivity, we have

$$\Delta\rho = \alpha_D \, \Delta T_f + \alpha_m \Delta T_m$$

Typical values for coefficients are (see [13])

$$\alpha_D = -2.5 \, \text{pcm/K}$$

$\alpha_m = -15 \, \text{pcm/K}$ at BOC (Beginning of Cycle).

$\alpha_m = -60 \, \text{pcm/K}$ at EOC (End of Cycle).

The explanation for different values of α_m at BOC and EOC is that the moderator contains 1500 ppm of Boron at BOC whereas it is pure water at EOC.

Based on a Joint Work with P. Reuss [14]

B. Mercier, *Simple Models for Understanding Nuclear Reactor Physics*,
https://doi.org/10.1007/978-3-031-98532-4_15

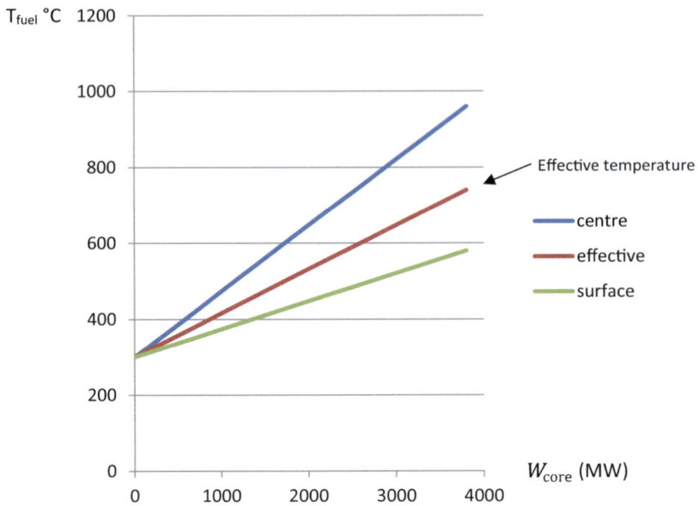

Fig. 15.1 Fuel temperature versus reactor power (PWR1300)

The (absolute) value given for α_D may look small at first sight, but the fuel temperature variations are much larger than the moderator temperature variations due to the weak conductivity of UO_2 (see Exercise 15.1).

Figure 15.1 shows the variation of fuel temperature with respect to reactor power in a PWR1300. Three curves are given: one at the center of the fuel pin, one at the surface and one for the effective temperature. The value we give for α_D is valid for the effective temperature variation.

Note that, in normal operations, the effective fuel temperature ranges from 300 °C (hot shutdown) to 750 °C (RP = Rated power), so that the Doppler effect is responsible for a negative reactivity of almost – 1000 pcm between zero power and rated power (3800 MW in the case of Fig. 15.1).

At the same time the average moderator temperature ranges from 297 °C at zero power to 306 °C at rated power, so the moderator effect is responsible for a negative reactivity that can be – 135 pcm at BOC to—540 pcm at EOC.

Exercise 15.1 (*Thermal analysis of a fuel pin*)
We assume that the fuel pin's radius is b and that it contains:

- Uranium oxide (UO_2) for $0 < r < a$.
- A Zirconium cladding for $a < r < b$.

We denote by λ_0 the UO_2 thermal conductivity (W/m/K) and by λ_1 the cladding thermal conductivity.

For simplicity, we assume that λ_0 and λ_1 do not depend on temperature.

1. Let P denote the thermal power (W/cm^3) released by nuclear reactions in the UO_2 fuel pellet. Give the mathematical relation existing between the thermal flux $\vec{F} = -\lambda_0 \overrightarrow{\nabla T}$ and P.
2. The temperature field being assumed to have cylindrical symmetry ($T = T(r)$), show that in cylindrical coordinates, only the radial component F_r of $\vec{F}$ along r does not vanish and that we have $F_r = -\lambda_0 \frac{dT}{dr}$.
3. Using the formula

$$\operatorname{div} \vec{F} = \frac{1}{r}\frac{\partial}{\partial r}(rF_r) + \frac{1}{r}\frac{\partial}{\partial \theta}F_\theta + \frac{\partial}{\partial z}F_z$$

deduce that

(1) $-\lambda_0 \frac{1}{r}\frac{d}{dr}\left(r\frac{dT}{dr}\right) = P$ for $0 < r < a$
(2) $-\lambda_1 \frac{1}{r}\frac{d}{dr}\left(r\frac{dT}{dr}\right) = 0$ for $a < r < b$

4. From now on, we call $T_0 = T_0(r)$ the restriction of T to the interval $(0, a)$ and in the same way $T_1 = T_1(r)$ the restriction of T to (a, b). Show that if we let $F_0 = -\lambda_0 \frac{d}{dr} T_0$ then

$$F_0 = \frac{P \cdot r}{2}$$

Deduce that if $P(r) = P > 0$ does not depend on r for $0 < r < a$.

$$T_0(r) = T_a + \frac{P\left(a^2 - r^2\right)}{4\,\lambda_0}$$

where $T_a = T_0(a)$ is the temperature just at the cladding.
5. Let $F_1 = -\lambda_1 \frac{d}{dr} T_1$, show that
   $F_1 = \frac{c}{r}$ where c is some constant to be determined.
   deduce that $T_1 = -\frac{1}{\lambda_1}(c\,\mathrm{Ln}\,r + d)$ where d is some other constant.
   We assume that the cladding temperature is equal to the (assumed to be known) water temperature in $r = b$, so that $T_1(b) = T_b$. deduce that $T_1(r) = T_b + \frac{c}{\lambda_1}\mathrm{Ln}\frac{b}{r}$.
6. Show that $c = \frac{P \cdot a^2}{2}$.
7. We assume that there is no contact thermal resistance at the fuel pellet/cladding interface. Show that

$$T_0(r) = \frac{P\left(a^2 - r^2\right)}{4\lambda_0} + \frac{P \cdot a^2}{2\lambda_1}\mathrm{Ln}\frac{b}{a} + T_b$$

Numerical application: Take $a = 0.475$ cm $b = 0.532$ cm, $\lambda_0 = 0.05$ W/cm/K, $\lambda_1 = 0.2$ W/cm/K, $P = 200$ W/cm^3, $T_b = 600$ K, compute T_a then T_c.

Plot the curve $r \to T(r)$. Explain what happens at the interface.■ *(solution p. 197)*.

Of course, in a nuclear reactor, temperature depends on power, and power depends on the number of fissions per second, then on the neutron flux Φ, then on the number of neutrons in the core.

In what follows, we shall denote by $n(t)$ the ratio $P(t)/P_N$ where P_N is the rated power and $P(t)$ power at time t.

We shall first ignore delayed neutrons.

As in Chap. 11, our first equation will be:

$$\frac{dn}{dt} = \frac{\rho}{\ell_0} n. \tag{15.1}$$

where ℓ_0 is the lifetime obtained when the reactor is critical (see Chap. 11)

When $n(t)$ increases, $P(t)$ increases, the core temperature increases. More precisely T_f and T_m increase but not at the same rate.

When $n(t)$ increases from 0 to 1, there is a negative reactivity injection equal to − 1260 pcm at BOC and − 1665 pcm at EOC.

Assuming that the negative reactivity injection depends linearly on the variation of power (the general case is considered in [13]) we then have

$$d\rho = -\alpha \, dn$$

with $\alpha = \left\{ \begin{matrix} 1260 \text{ pcm at BOC} \\ 1665 \text{ pcm at EOC} \end{matrix} \right\}$

Our second equation is then:

$$\frac{d\rho}{dt} = -\alpha \frac{dn}{dt} = -\frac{\alpha}{\ell_0} \rho n \tag{15.2}$$

We then have to solve the nonlinear system

$$\frac{d}{dt}\begin{pmatrix} \rho \\ n \end{pmatrix} = \begin{pmatrix} -\frac{\alpha}{\ell_0}\rho n \\ \frac{\rho}{\ell_0} n \end{pmatrix} \tag{15.3}$$

Let $\begin{pmatrix} \rho_0 \\ n_0 \end{pmatrix}$ be the starting point (at time $t = 0$). We assume $n_0 > 0$ but ρ_0 can be positive or negative.

We shall prove that $n(t) \to n_\infty$ *as* $t \to \infty$ where $n_\infty = \max\left(0, n_0 + \frac{\rho_0}{\alpha}\right)$ and that $\rho(t) \to 0$ if $n_\infty > 0$.

Remarks 15.1

1. In our model, we assume that the reactor was in the critical state, at power n_0 for $t < 0$, when at time $t = 0$ we inject reactivity ρ_0, for example by raising or lowering the control rods, and that we then let the reactor evolve naturally without moving the control rods.

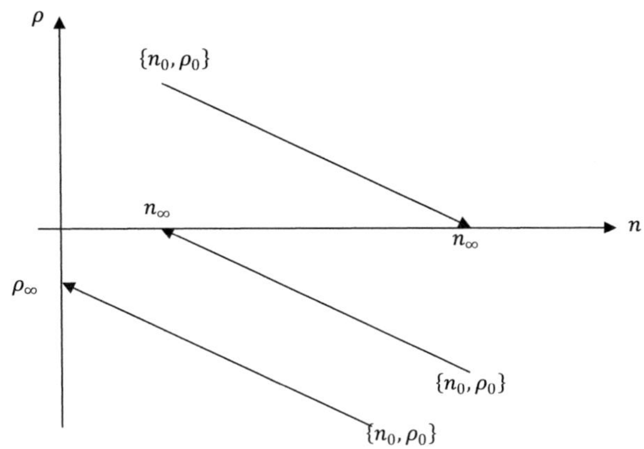

Fig. 15.2 Evolution of the solution in plane $\{n, \rho\}$

2. We easily check that

$$\frac{d\rho}{dn} = \frac{d\rho}{dt} \Big/ \frac{dn}{dt} = \frac{-\frac{\alpha}{\tau}\rho n}{\frac{1}{\tau}\rho n} = -\alpha$$

   We deduce that the curves $t \to \{n(t), \rho(t)\}$ are describing straight lines with slope $-\alpha$ in the plane $\{n, \rho\}$ hence:

$$\rho(t) = \rho_0 + \alpha(n_0 - n(t)) = \alpha(n_\infty - n(t)) \tag{15.4}$$

3. Dividing the first Eq. (15.3) by ρ and the second one by n we get:

$$\frac{d}{dt}\begin{pmatrix} \ln|\rho| \\ \ln n \end{pmatrix} = \begin{pmatrix} -\frac{\alpha}{\ell_0} n \\ \frac{\rho}{\ell_0} \end{pmatrix} \tag{15.5}$$

   We deduce that if we start from the domain $\{\rho > 0, n > 0\}$, we stay there.
   The same for the domain $\{\rho < 0, n > 0\}$.
   In domain $\{\rho > 0, n > 0\}$, $\rho(t)$ decreases and $n(t)$ increases.
   In domain $\{\rho < 0, n > 0\}$, $\rho(t)$ increases and $n(t)$ decreases. (see Fig. 15.2).
4. We see on Fig. 15.2 that 3 cases are possible:
   1. $\rho_0 > 0$: then, the straight line $\rho = \rho_0 + \alpha(n_0 - n)$ cuts the $n-$ axis in $n_\infty = n_0 + \frac{\rho_0}{\alpha}$.
   2. $\rho_0 < 0$, and $n_0 + \frac{\rho_0}{\alpha} > 0$ then the straight line $\rho = \rho_0 + \alpha(n_0 - n)$ cuts the $n-$ axis in $n_\infty = n_0 + \frac{\rho_0}{\alpha}$.
   3. $\rho_0 < 0$, and $n_0 + \frac{\rho_0}{\alpha} < 0$ then the straight line $\rho = \rho_0 + \alpha(n_0 - n)$ cuts the $\rho - axis$ in $\rho_\infty = \alpha n_0 + \rho_0$.

We shall prove that in cases 1 and 2, $\rho(t) \to 0$ and $n(t) \to n_\infty$, whereas in case 3.

$\rho(t) \to \rho_\infty$ and $n(t) \to 0$.

Solution of System (15.3) **(See [14])**

We proceed by separation of variables.

We use (15.4) to eliminate ρ.

We obtain

$$\frac{\mathrm{d}n}{\mathrm{d}t} = \frac{\alpha}{\ell_0} n(n_\infty - n(t))$$

i-e

$$\frac{\mathrm{d}n}{n(n_\infty - n)} = \alpha \frac{\mathrm{d}t}{\ell_0}$$

$$\frac{\mathrm{d}n}{n_\infty}\left[\frac{n_\infty - n + n}{n[n_\infty - n]}\right] = \frac{\mathrm{d}n}{n_\infty}\left[\frac{1}{n} + \frac{1}{n_\infty - n}\right] = \alpha \frac{\mathrm{d}t}{\ell_0}$$

Thus, by integration:

$$\frac{1}{n_\infty}[\ln n - \ln|n_\infty - n|] = \alpha \frac{t}{\ell_0} + B$$

$$-\frac{1}{n_\infty} \ln \frac{|n_\infty - n|}{n} = \alpha \frac{t}{\ell_0} + B$$

To find the value of B, we apply this relation at $t = 0$:

$$B = -\frac{1}{n_\infty} \ln \frac{|n_\infty - n_0|}{n_0}.$$

We obtain:

$$\alpha \frac{t}{\ell_0} = \frac{1}{n_\infty} \ln \frac{|n_\infty - n_0|}{n_0} \frac{n}{|n_\infty - n|}$$

$$\frac{|n_\infty - n|}{n} \frac{n_0}{|n_\infty - n_0|} = \exp\left(-\alpha n_\infty \frac{t}{\ell_0}\right)$$

$$n|n_\infty - n_0| \exp\left(-\alpha n_\infty \frac{t}{\ell_0}\right) = n_0|n_\infty - n|$$

In the case where $n_\infty > n > n_0$ we find

$$n\left[n_0 + (n_\infty - n_0) \exp\left(-\alpha n_\infty \frac{t}{\ell_0}\right)\right] = n_0 n_\infty$$

$$n(t) = \frac{n_0 n_\infty}{n_0 + (n_\infty - n_0)\exp\left(-\alpha n_\infty \frac{t}{\ell_0}\right)} \tag{15.6}$$

We see that $n(t) \to n_\infty$ hence $\rho(t) \to 0$.

In the case where $0 < n_\infty < n < n_0$

$$n\left[-n_0 + (n_0 - n_\infty)\exp\left(-\alpha n_\infty \frac{t}{\ell_0}\right)\right] = -n_0 n_\infty$$

And we get the same result.

We get $n(t) \to n_\infty$ hence $\rho(t) \to 0$.

In the case $n_\infty < 0 < n < n_0$ formula (15.6) shows that $n(t) \to 0$. We deduce that $\rho(t) \to \alpha n_\infty$.

Numerical Values for a PWR

Note that if $n_0 = 0$ and $\rho_0 > 0$, then $n_\infty = \rho_0/\alpha$. So to go from critical state at zero power to critical state at rated power we need to inject reactivity $\rho_0 = \alpha$. So α is called the "power fault". This varies between the BOC and the EOC.

If we start from $n_0 = 0.2$ and $\rho_0 = 100$ pcm, we reach $n_\infty = 0.28$ at BOC and $n_\infty = 0.26$ at EOC.

If we start from $n_0 = 0.2$ and $\rho_0 = -100$ pcm, we reach $n_\infty = 0.12$ at BOC (0.14 at EOC).

Remarks 15.2

1. Obviously, we should not generate suddenly a positive reactivity of + 1300 pcm, otherwise we would get prompt criticity and the core could get damaged: it is better to go step by step.
2. For a reactor where temperature effects are not significant, to increase power, first raise the bars, then lower them again, once you've reached the desired power. When temperature effects are significant (example PWR), this is not necessary if the bars have been raised to the right level…
3. Rather than speaking of "temperature effects" some authors speak of "counter-reactions". Mathematically we should say that they are non-linear effects.

Taking Delayed Neutrons into Account

We start from the 6 groups model introduced in Chap. 11:

$$\frac{dn}{dt} = \frac{\rho - \beta}{\ell_0} n + \sum_i \lambda_i c_i \tag{15.7}$$

$$\frac{dc_i}{dt} = \frac{\beta_i}{\ell_0} n - \lambda_i c_i \quad 1 \le i \le 6 \tag{15.8}$$

To put us in a situation comparable to the above, we have normalized these equations by systematically dividing them by the number N of neutrons that are in the

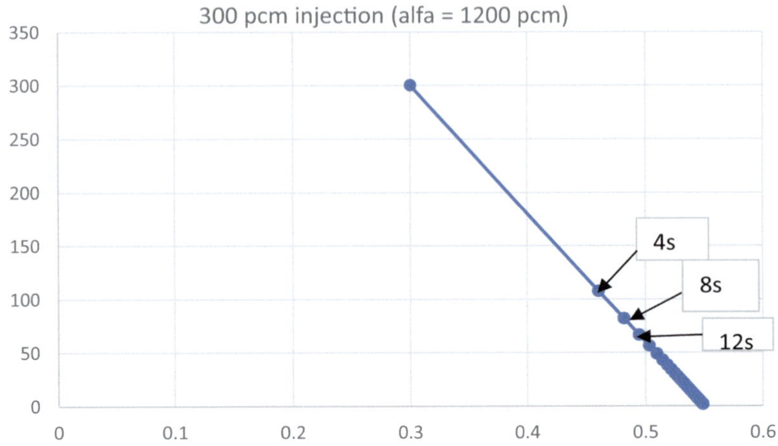

Fig. 15.3 Graph giving $\rho(t)$ versus $n(t)$ in the case with delayed neutrons

core at rated power so $n = n(t)$ is the normalized power and $c_i = c_i(t)$ the number of precursors also normalized by N, λ_i the associated decay constant, ℓ_0 the neutron lifetime and $\beta = \sum_i \beta_i$ the fraction of delayed neutrons.

To take temperature effects into account, we proceed as above: we decide that ρ is the 8th variable and we complete this 7×7 differential system with an 8th equation:

$$\frac{\mathrm{d}\rho}{\mathrm{d}t} = -\alpha \frac{\mathrm{d}n}{\mathrm{d}t} \tag{15.9}$$

We complement with initial conditions:

$$n(0) = n_0;\ c_i(0) = c_i^0;\ \rho(0) = \rho_0$$

We can expect that $\rho(t) \to 0$ as $t \to \infty$ and also (taking $\frac{\mathrm{d}n}{\mathrm{d}t} = \frac{\mathrm{d}c_i}{\mathrm{d}t} = 0$ in (15.7) and (15.8)) that

$$n(t) \to n_\infty = n_0 + \frac{\rho_0}{\alpha}$$

$$c_i(t) \to \frac{\beta_i}{\ell_0} n_\infty$$

Of course, the kinetics will not be the same as in (15.3) but, if we start from the same initial power n_0 with the same reactivity injection ρ_0, the final power n_∞ will be the same.

In Fig. 15.3 we give the result of a simulation for the model with delayed neutrons with PWR data.

We can see that the kinetics is rather slow.

Chapter 16
Introduction to PWR System Simulation

Notations

PN	Nominal power
ρ	Core reactivity (pcm)
x	Normalized number of neutrons or normalized core power (equal to 1 at PN)
y	Normalized steam valve opening or normalized secondary power released (equal to 1 at PN)
c	Normalized number of precursors
α_D	Doppler coefficient (pcm/°)
α_m	Moderator coefficient (pcm/°)
mc	Fuel's thermal inertia (MJ/°)
a	Nominal power (PN) of the core (MW)
$h \cdot s$	Fuel $\rightarrow$ moderator heat transfer coefficient (MW/°)
T_c	Fuel average temperature (°C)
T_m	Moderator average temperature (°C)
$M \cdot C_p$	Primary circuit (water + structure) thermal inertia (MJ/°)
$H \cdot S$	Primary $\rightarrow$ secondary heat transfer coefficient (MW/°)
T_v	(saturating) steam temperature (secondary side) (°C)
l	Neutrons' lifetime
λ	Decay constant of precursors emitting delayed neutrons (s^{-1})
β	Delayed neutrons fraction (pcm)
BOC	Beginning of cycle
EOC	End of cycle

A PWR is obviously a complex system. Full scale simulators involve the simulation of many coupled phenomena, in particular to simulate accidents. In this chapter we

(Chapter based on a joint work with Bruno Tarride, CEA/INSTN [17])

B. Mercier, *Simple Models for Understanding Nuclear Reactor Physics*,
https://doi.org/10.1007/978-3-031-98532-4_16

shall restrict ourselves to normal PWR operations where the primary circuit remains at the nominal pressure (155 bars).

With a reasonably simple 5-equations model, we show that a PWR behaves differently, depending on whether it operates in the follower secondary mode or in the priority secondary mode.

16.1 Equations of Our Model

In a PWR, neutron-generated power $a \cdot x$ is released within the fuel and discharged to the moderator (= coolant = water).

If h is the heat transfer coefficient and s the exchange surface area, the power exchanged between fuel and moderator is expressed by $h \cdot s(T_c - T_m)$.

Similarly, the power transferred from primary to secondary is expressed by $H \cdot S(T_m - T_v)$.

Our ultra-simplified PWR model can then be described by the following diagram (cf [15]):

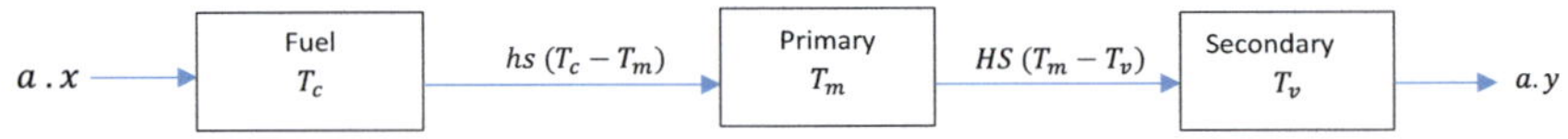

Energy balances can be written in the classic form: accumulation = input − output.

We deduce:

$$\text{Fuel:} \quad m \cdot c \frac{d}{dt} T_c = a \cdot x - h \cdot s(T_c - T_m)$$

$$\text{Primary:} \quad M \cdot C_p \frac{d}{dt} T_m = h \cdot s(T_c - T_m) - H \cdot S(T_m - T_v)$$

To find out how the core power $a \cdot x$ evolves, we need to evaluate its reactivity ρ.

This can vary depending on the reactivity devices (control rods or soluble boron) or on the negative reactivity effects involving neutronics and thermal-hydraulics:

- The Doppler effect $\alpha_D = \frac{\partial \rho}{\partial T_c}$
- The moderator effect $\alpha_m = \frac{\partial \rho}{\partial T_m}$

In the absence of manual control on the reactivity devices, the reactivity balance is as follows:

$$\frac{d\rho}{dt} = \alpha_D \frac{d}{dt} T_c + \alpha_m \frac{d}{dt} T_m \tag{16.1}$$

To link the core reactivity and the number of neutrons, proportional to core power, we use the balance equations for 1-group precursor kinetics at the origin of delayed neutrons:

$$\text{neutrons:} \quad \frac{d}{dt}x = \frac{\rho - \beta}{l}x + \lambda c$$

$$\text{precursors:} \quad \frac{d}{dt}c = \frac{\beta}{l}x - \lambda c$$

So, we obtain the differential system:

$$\frac{d\rho}{dt} = \frac{\alpha_D}{m \cdot c}(a \cdot x - hs(T_c - T_m)) + \frac{\alpha_m}{MC_p}(hs(T_c - T_m) - HS(T_m - T_v)) \quad (16.2)$$

$$\frac{d}{dt}x = \frac{\rho - \beta}{l}x + \lambda c \quad (16.3)$$

$$\frac{d}{dt}c = \frac{\beta}{l}x - \lambda c \quad (16.4)$$

$$m \cdot c\frac{d}{dt}T_c = a \cdot x - hs(T_c - T_m) \quad (16.5)$$

$$MC_p\frac{d}{dt}T_m = hs(T_c - T_m) - HS(T_m - T_v) \quad (16.6)$$

At this stage, we note that the reactor can operate following two modes:

(a) At 5%PN, the turbine is not yet coupled; steam is discharged to the turbine bypass system, where the steam valve regulates the secondary pressure to a set point.

With the secondary at saturation, this gives us the following complementary equation:

$$T_v = \text{constant} \quad (16.7a)$$

We say it's a follower secondary mode because the valve, in order to fix the steam pressure, will have its position y adapted so as to evacuate the power produced at the primary.

(b) Above 5%PN, the steam is discharged to the turbine, via a valve whose position is directly linked to the power imposed by the operator on the turbine-generator set. As a 1st approximation, the secondary power can be considered to be, at any given time:

$$H \cdot S(T_m - T_v) = a \cdot y, \quad (16.7b)$$

y being given.

The secondary is said to have priority, because it is fixed from the initial moment.

The implications of this situation will be seen later.

16.2 Steady State

When time $t \to \infty$ we obtain a steady state such that:

$$a \cdot x = hs(T_c - T_m) = HS(T_m - T_v) = a \cdot y, \text{ so that } x = y$$

$$\rho = 0 \quad \text{and} \quad \frac{\beta}{l}x = \lambda c$$

16.3 Numerical Data

For a 4 loops PWR whose core power is 3800 MW:

From [2], as far as neutron kinetics is concerned we shall take:

- $l = 25\,\mu\text{s}$,
- β delayed neutrons fraction ($\beta \approx 0.6\%$); $\lambda = 0.08\,\text{s}^{-1}$

From reference [15] we can take:

- Negative-reactivity effects: Doppler $\alpha_D = -3\,\text{pcm}/°$

  Moderator α_m goes from $-10\,\text{pcm}/°$(BOC) to $-60\,\text{pcm}/°$(EOC).

Thermohydraulic data:

- fuel's thermal inertia: m·c = 40 MJ/°.
- fuel → moderator heat transfer coefficient: h·s = 7 MW/°.
- primary circuit (water + structure) thermal inertia: $M \cdot C_p = 3000$ MJ/°
- primary → secondary heat transfer coefficient (4 loops): $H \cdot S = 160$ MW/°.

16.4 Time Discretization

For (16.3) and (16.4) we shall use the implicit backward difference scheme introduced in Chap. 11.

Like in Chap. 11, we shall introduce a time step τ; given the values ρ, x, c, T_c, T_m at time t, we denote the new values at time $t + \tau$ by $\rho^*, x^*, c^*, T_c^*, T_m^*$. We require that

$$\frac{x^* - x}{\tau} = \frac{\rho - \beta}{l}x^* + \lambda c^*$$

$$\frac{c^* - c}{\tau} = \frac{\beta}{l}x^* - \lambda c^*$$

This is a 2 × 2 linear system for $\{x^*, c^*\}$ that we solve as in chap. 11.
In addition, we require that

$$m \cdot c \cdot \frac{T_c^* - T_c}{\tau} = a \cdot x^* - hs(T_c - T_m) \quad \text{for (16.5)} \tag{16.8}$$

$$MC_p \frac{T_m^* - T_m}{\tau} = hs(T_c - T_m) - HS(T_m - T_v) \quad \text{for (16.6)} \tag{16.9}$$

$$\rho^* - \rho = \alpha_D\left(T_c^* - T_c\right) + \alpha_m\left(T_m^* - T_m\right) \quad \text{for (16.2)} \tag{16.10}$$

16.5 Performed Simulations

All simulated transients start from a stationary state indicated by superscript ^{0}.

16.5.1 Follower Secondary

Case 1: Initial state 1% PN—100 pcm positive reactivity injected by control rods extraction

The core is at a BOC and we assume a reactivity step $\Delta\rho = +100$ pcm at time $t = 0$.
The results obtained with $\tau = 1$ s are given in Figs. 16.1, 16.2 and 16.3.
As $t \to \infty$ we get a new (critical) steady state indicated by a superscript ^{1}.
We obtain the following differences between the final state and the initial state:

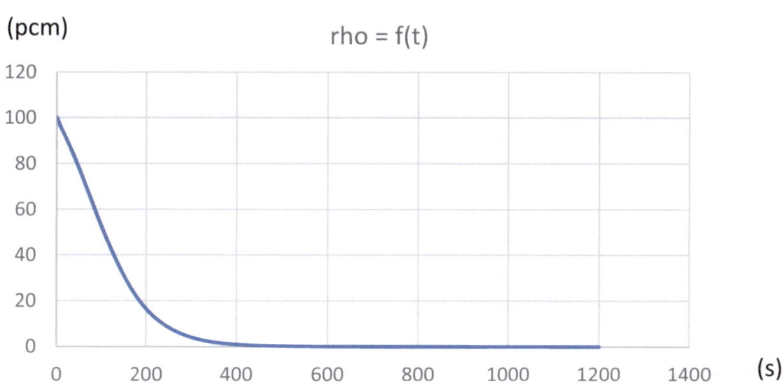

Fig. 16.1 Follower secondary: reactivity evolution

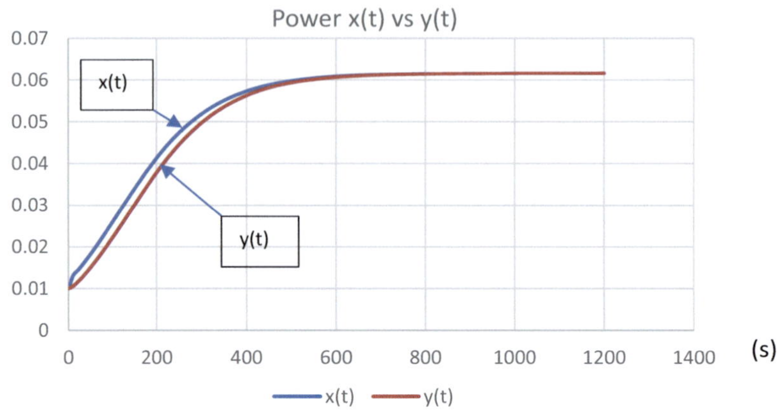

Fig. 16.2 Follower secondary: power coefficients evolution

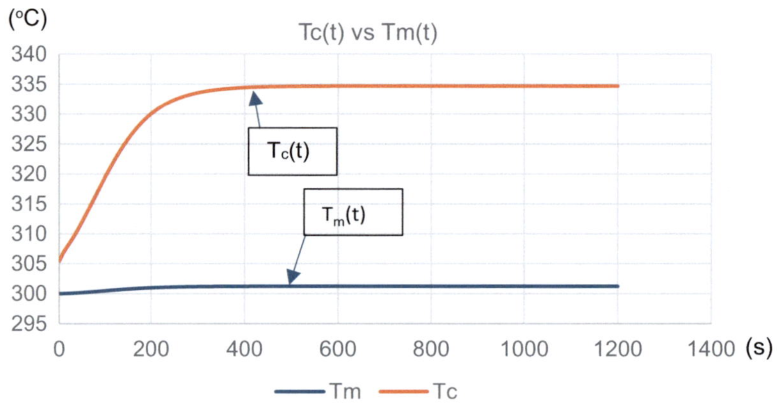

Fig. 16.3 Follower secondary: fuel and moderator temperatures evolution

$$\Delta W = a \cdot x^1 - a \cdot x^0 = a \cdot y^1 - a \cdot y$$
$$\Delta T_c = T_c^1 - T_c^0$$
$$\Delta T_m = T_m^1 - T_m^0$$
$$\Delta T_v = T_v^1 - T_v^0$$

Now, in the follower secondary mode, we have $\Delta T_v = 0$ so that

$$\Delta T_m = \frac{\Delta W}{HS}$$
$$\Delta T_c = \Delta T_m + \Delta W / hs$$

Integrating (16.1) from $t = 0$ to $t = \infty$ we get

$$0 = \Delta\rho + \alpha_D \Delta T_c + \alpha_m \Delta T_m$$

so that:

$$\alpha_D \left(\frac{\Delta W}{HS} + \frac{\Delta W}{hs} \right) + \alpha_m \frac{\Delta W}{HS} = -\Delta\rho$$

Let $\theta = \frac{H.S}{h.s}$, we get

$$\Delta W = -\frac{H.S.\Delta\rho}{\alpha_D(1+\theta) + \alpha_m}$$

With the selected numerical data, we obtain $\Delta W = 196$ MW, that is 5.1%PN.

We check on Fig. 16.2 that, with a 100 pcm reactivity insertion, the core power and the secondary power both increase from 1 to 6.1%PN.

We also have

$$\alpha_D \Delta T_c = \frac{(1+\theta)\alpha_D}{\alpha_D(1+\theta)+\alpha_m} \Delta\rho \quad \text{and} \quad \alpha_m \Delta T_m = \frac{\alpha_m}{\alpha_D(1+\theta)+\alpha_m} \Delta\rho$$

so that the negative reactivity effects are distributed with 88% for the Doppler effect and 12% for the moderator effect. (At EOC, it would be 54% for the Doppler effect and 46% for the moderator effect).

Lesson 1: This behavior was expected; we can see that we don't need to play with the absorbers to stabilize the core; the core naturally becomes just critical again, thanks to the effect of moderator and, above all, Doppler feedback.

Our model clearly shows that PWRs are stable.

16.5.2 Priority Secondary

Case 2: Initial state 50% PN—100 pcm reactivity injection via control rods extraction.

The following simulation is performed twice, first with $\alpha_m = -10\,\text{pcm}/^\circ$ (BOC) then with

$$\alpha_m = -60\,\text{pcm}/^\circ(\text{EOC})$$

Figure 16.4 shows that reactivity oscillates slightly more at EOC than at BOC.

Figure 16.5 shows that with the same reactivity injection the effect on the moderator temperature is more sensitive at BOC.

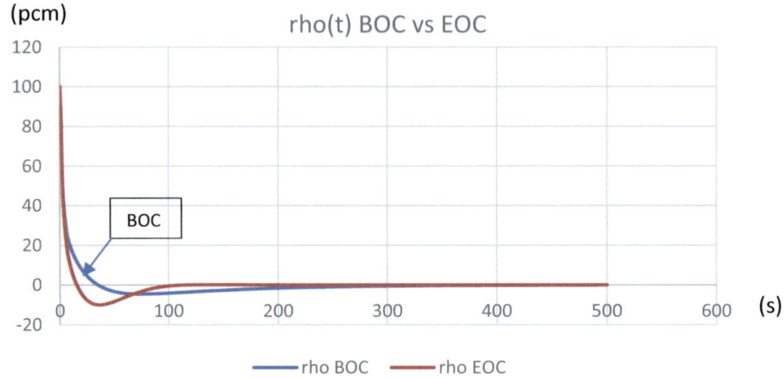

Fig. 16.4 Case 2. Reactivity evolution BOC versus EOC

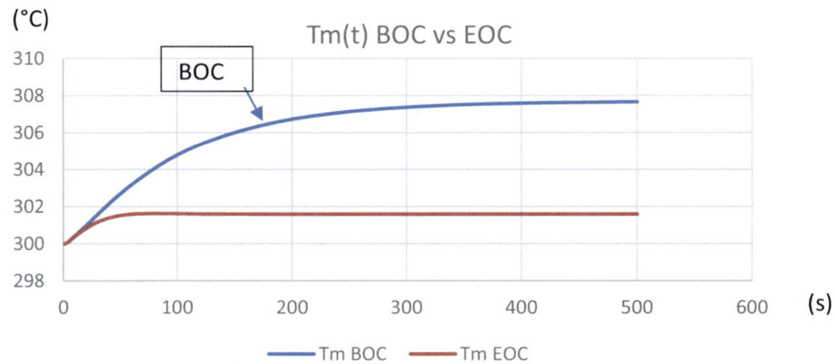

Fig. 16.5 Case 2. Moderator temperature evolution BOC versus EOC

Figure 16.6 confirms that the secondary power is constant while the core power has a significant peak in the first seconds of our simulation.

We still introduce a Δ symbol to indicate the difference between the final state and the initial state.

We have here $\Delta W = 0$, hence

$$\Delta T_v = \Delta T_m = \Delta T_c$$

The reactivity balance gives again

$$\Delta\rho + \alpha_D \Delta T_c + \alpha_m \Delta T_m = 0$$

so that $\Delta T_m = -\frac{\Delta\rho}{\alpha_D + \alpha_m}$

With the selected data, we get, $\Delta T_m = +\,7.7$ °C at BOC and $\Delta T_m = +\,1.6$ °C at EOC, which is consistent with Fig. 16.5.

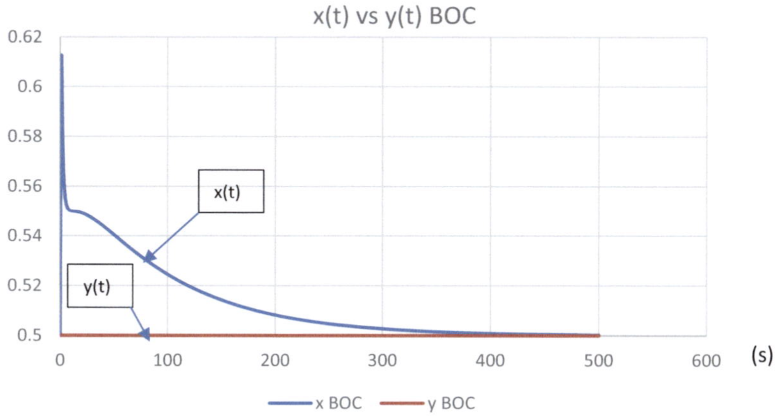

Fig. 16.6 Case 2. Core power versus secondary power at BOC

Lesson 2: this case is actually rather counter-intuitive: inserting the same reactivity as in the previous case doesn't result in a permanent increase in power; in fact, the core power is brought back, by Doppler/moderator feedback, to the power demand at the secondary, which has remained unchanged during the transient.

In other words, the reactivity insertion has the effect of heating the primary water: **we deduce from this fact that a possible function of the control rods is, in the priority secondary mode, to adjust the moderator temperature T_m!**

Case 3: Initial state 50% PN—increase in secondary power, priority secondary mode

This case will be studied both at BOC and EOC.

On the basis of the previous lesson, we shall open the turbine inlet valve from 0.5 to 0.6, starting from the same critical state $x^0 = 0.5$.

We can see that the core is following the secondary demand although with some delay. This is quite practical for the operator, who can impose his/her conditions on the turbine-generator set to meet network requirements… and the core follows through the effect of Doppler/moderator feedback, with no need to call on the control rods, Figs. 16.7, 16.8, 16.9 and 16.10

The downside is the primary/secondary cooling effect, which can be evaluated as follows.

Using the same notation as above, between the initial and final states, we get in this case:

$$\Delta W = hs(\Delta T_c - \Delta T_m) \text{ hence } \Delta T_c = \Delta T_m + \frac{\Delta W}{hs}$$

$$\alpha_D \Delta T_c + \alpha_m \Delta T_m = 0 \text{ hence } \alpha_D\left(\Delta T_m + \frac{\Delta W}{hs}\right) + \alpha_m \Delta T_m = 0$$

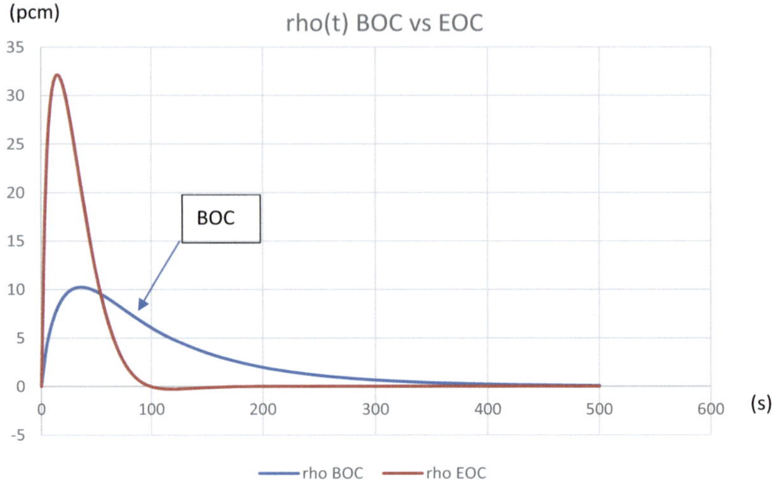

Fig. 16.7 Case 3: reactivity evolution at BOC versus EOC

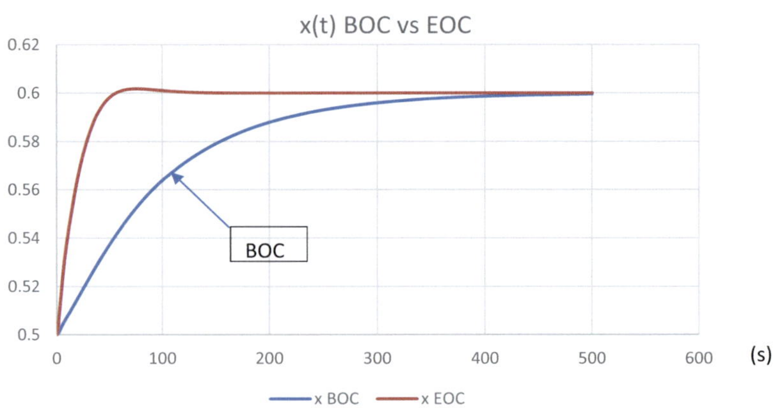

Fig. 16.8 Case 3: Core power evolution at BOC versus EOC

$$\Delta T_m = -\frac{\alpha_D}{\alpha_D + \alpha_m}\frac{\Delta W}{hs}$$

This gives for the moderator temperature − 12.5° at BOC and − 2.6° at EOC. Finally, if the fuel heats up, the primary and secondary cool down.

This relationship of decay of T_{m} and T_{v} as a function of power is called the natural temperature program (linked to feedback effects alone).

Lesson 3: Such a deterioration in steam quality (T_{v}, P_{v}) during ramp-up will be accompanied by a drop in the plant's efficiency (steam is the "hot source" in Carnot's sense): heat is converted less effectively into electricity, which is penalizing.

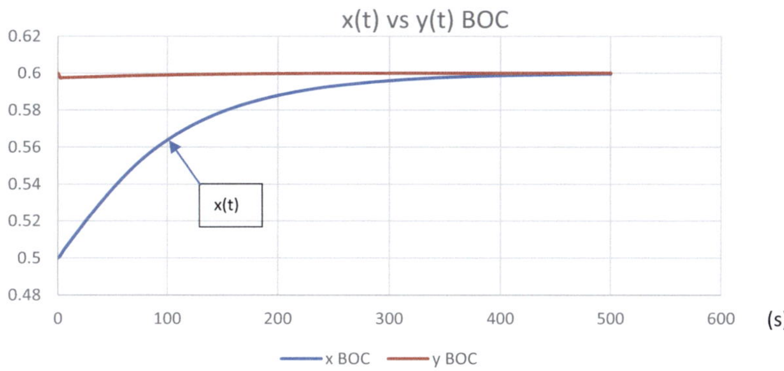

Fig. 16.9 Case 3: core power versus secondary power at BOC

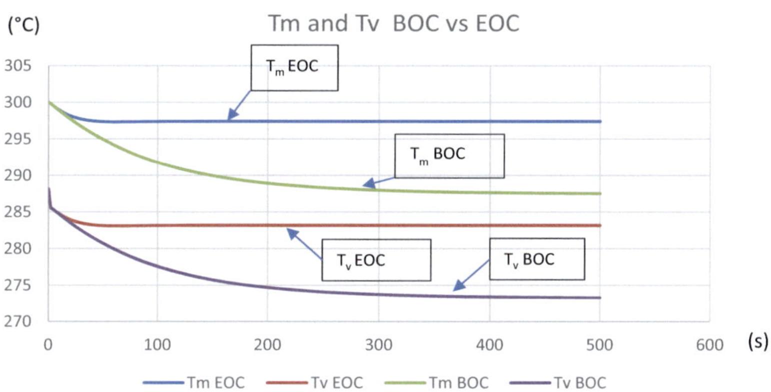

Fig. 16.10 Case 3: Moderator and steam temperature evolution at BOC versus EOC

This loss of efficiency is not acceptable to an electrician, who will prefer to impose an "artificial" temperature program.

To do this, the operator imposes a setpoint program $T_{\mathrm{m}} = f(y)$ and the use of a group of control absorbers to adjust T_{m} to its setpoint value (see case 2).

Lesson 4: Case 3 highlights the "complex" nature of a PWR, as well as a particular family of reactivity accidents, such as the steamline break.

For example, a steamline break in the machine room, close to the turbine-generator set, will result in a secondary power demand that will be transmitted to the primary system; the induced primary cooling inserts reactivity into the core through moderating feedback.

Our model can therefore be used to raise awareness of this accident, and in particular of the penalizing conditions to be involved in the reactor safety study. For safety studies $\frac{\mathrm{d}\rho}{\mathrm{d}t}$ should be maximized, i-e.

- EOC, with a strongly negative α_m,
- new steam generators (clean tubes maximizing the heat transfer coefficient HS) and thermal inertia $M \cdot C_p$ reduced).

Conclusion for Chapter 16

This very simple model of a PWR gives results close to those obtained on a simulator like the one developed in France by CORYS. We also refer to the IAEA simulator [16].

Operated with just three transients, it is quite rich in information:

(1) Doppler and moderator feedback always bring reactivity back to 0; as a result, PWRs are neutron-stable, a factor that is highly conducive to compliance with the primary safety function of chain reaction control.
(2) There are two operating modes on a PWR: follower secondary and priority secondary; for the same initiating event, the general behavior will not be the same; some behaviors are even counter-intuitive, which is a property of complex systems, characterized by interactions between sub-parts (here core/primary/secondary) and feedbacks (here Doppler/moderator).
(3) Priority turbine: a power increase must be made by a secondary power call; the core is then a follower due to the feedback effect; the consequence is a natural drop in T_{m} and $\mathrm{T_v}$; to avoid degrading the plant's efficiency, operators prefer to regulate T_{m} with a dedicated control rods group, providing the necessary reactivity, and follow the target given in Fig. 15.1.
(4) The involved physics implies that a steam line break can be the cause of a reactivity accident; the model helps to understand the transient, as well as the main penalizing hypotheses for its study.

Chapter 17
Xenon Oscillations (Based on a Joint Work with Zeng Ziliang, Chen Liyi, Shao Nuoya [18])

The Xenon effect is not just an overall effect on core reactivity: sinking the control rods can induce axial oscillation between the bottom and top of the core.

Oscillations can develop only if the core is large enough (compared to the diffusion length), which is the case for a PWR since the core is about 4 m high (and the diffusion length is around 2 cm).

In this chapter, we shall study such oscillations and explain how to control them.

Figure 17.1 (from [18]) shows an example of oscillation due to some initial perturbation: for $t < 12$ h there is some xenon poisoning in the upper part of the core and the neutron flux is higher in the lower part. The result of a lower neutron flux in the upper part is to reduce the production of ^{135}I and then ^{135}Xe. The Xenon poisoning reduces also by the effect of xenon β-decay.

At the same time, the higher neutron flux in the lower part produces more ^{135}I and then ^{135}Xe.

Xenon poisoning of the lower part is clear at $t = 24.7$ h. So this is how the oscillation begins.

To control such oscillations, as explained in Sect. 17.6, it is necessary to have two independent ways of adjusting the core reactivity.

This is the case on a PWR where one can regulate the control bars *and* the boron concentration in the moderator.

The *axial offset*, that evaluates the power imbalance between the upper and the lower part of the core, is a parameter to follow carefully for oscillations control.

This was not possible at low power on the RBMK and this is one element explaining the Chernobyl accident.

In chaps. 1–15, whenever possible, we have used analytical solutions to our simple models. This is not possible for Xenon oscillations. Our choice is to use 1D finite difference models.

Such models are easily implemented in Python with Numpy.

B. Mercier, *Simple Models for Understanding Nuclear Reactor Physics*,
https://doi.org/10.1007/978-3-031-98532-4_17

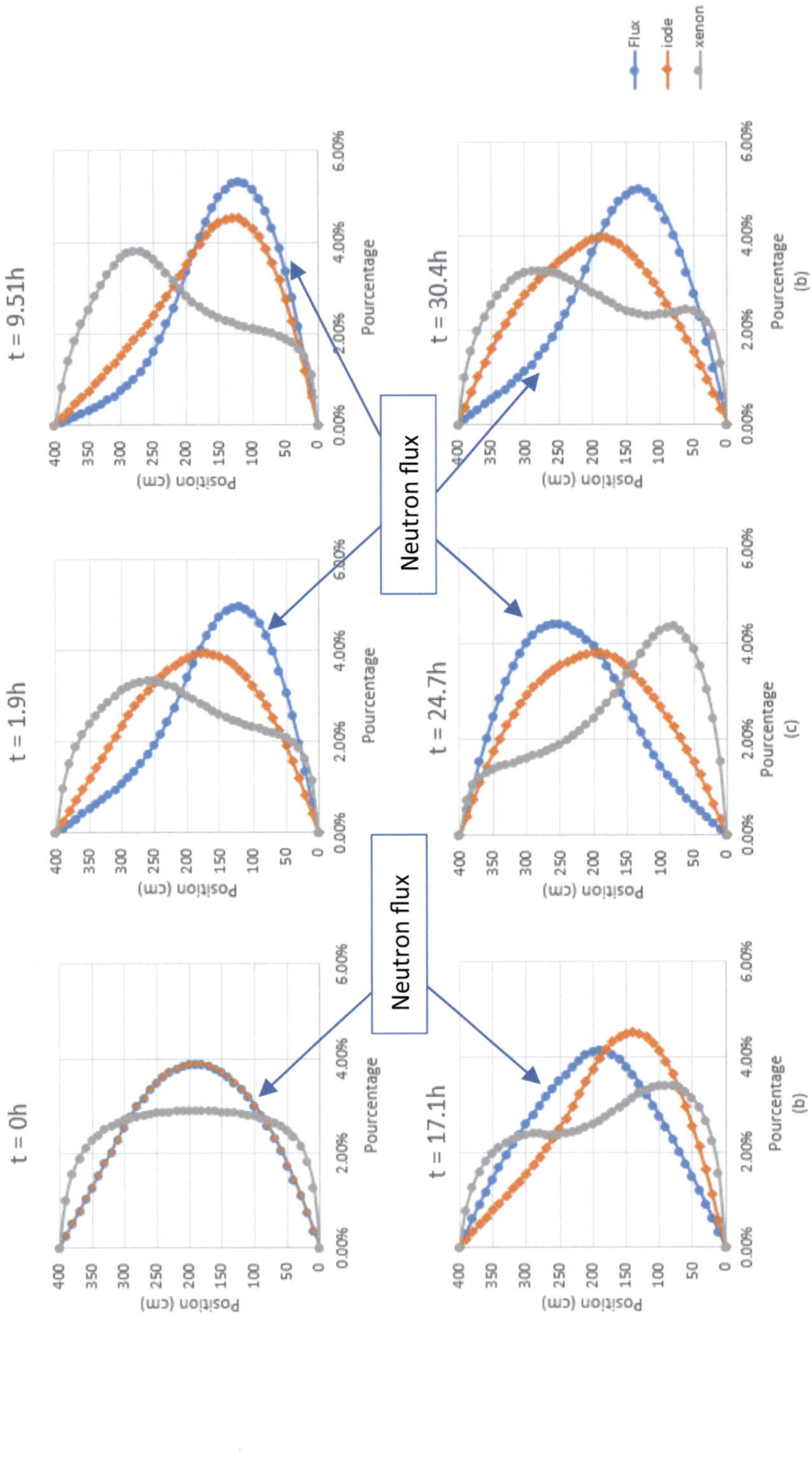

Fig. 17.1 Xenon oscillation development (from [18])

17.1 Implementation of the Fourier Boundary Condition

We start with the one group model in 1D, with the Fourier boundary condition (8.11)

$$-M^2\Phi'' + \Phi = k_\infty \Phi \quad \text{for } 0 \le z \le H \tag{17.1}$$

$$\Phi(0) - 2D\Phi'(0) = 0 \tag{17.2}$$

$$\Phi(H) + 2D\Phi'(H) = 0 \tag{17.3}$$

To set up our finite difference scheme, we shall use the variational approach. Let V be the space of functions, the square of which is integrable on (O, H) as well as their first derivative.

Let $\Psi \in V$ be an arbitrary test function: we have

$$\int_0^H \left(-M^2\Phi'' + \Phi - k_\infty\Phi\right)\Psi \mathrm{d}z = 0$$

so that by integration by parts, we get

$$M^2\int_0^H \Phi'\Psi' \mathrm{d}z - M^2\left[\Phi'\Psi\right]_0^H + \int_0^H (\Phi - k_\infty\Phi)\Psi \mathrm{d}z = 0$$

And, by using (17.2) and (17.3):

$$M^2\int_0^H \Phi'\Psi' \mathrm{d}z + \frac{M^2}{2D}(\Phi(0)\Psi(0) + \Phi(H)\Psi(H)) + \int_0^H (\Phi - k_\infty\Phi)\Psi \mathrm{d}z = 0$$

Let $h = H/N (\Phi_i)_{i=0,\ldots,N}$ and $(\Psi_i)_{i=0,\ldots,N}$ be finite difference approximations of Φ and Ψ we require

$$M^2\sum_0^{N-1} \frac{\Phi_{i+1} - \Phi_i}{h}\frac{\Psi_{i+1} - \Psi_i}{h} h + \frac{M^2}{2D}(\Phi_0\Psi_0 + \Phi_N\Psi_N) + \sum_0^{N-1}\frac{1 - k_\infty}{2}(\Phi_i\Psi_i + \Phi_{i+1}\Psi_{i+1})h = 0$$

Since this must be true for all $(\Psi_i)_{i=0,\ldots,N}$ we get

$$M^2\left(\frac{2}{h^2}+\frac{1}{Dh}\right)\Phi_0 - \frac{2M^2}{h^2}\Phi_1 + \Phi_0 = k_\infty \Phi_0$$

$$\frac{M^2}{h^2}(-\Phi_{i-1} + 2\Phi_i - \Phi_{i+1}) + \Phi_i = k_\infty \Phi_i \ \text{ for } i = 1, \ldots, N-1$$

$$-\frac{2M^2}{h^2}\Phi_{N-1} + M^2\left(\frac{2}{h^2}+\frac{1}{Dh}\right)\Phi_N + \Phi_N = k_\infty \Phi_N$$

which means that $k_\infty - 1$ is the smallest eigenvalue of the tridiagonal matrix

$$A = M^2\begin{pmatrix} \frac{2}{h^2}+\frac{1}{Dh} & -\frac{2}{h^2} & & & \\ \cdots & \cdots\ \cdots & & \cdots\ \cdots & \\ & -\frac{1}{h^2}\ \ \frac{2}{h^2} & & -\frac{1}{h^2} & \\ \cdots\cdots\cdots & & & \cdots & \cdots \\ & & & -\frac{2}{h^2} & \frac{2}{h^2}+\frac{1}{Dh} \end{pmatrix}$$

(even though this matrix is non-symmetric, we can easily prove it has real eigenvalues).

On Fig. 17.2, we show the fundamental eigenmode we get with $H = 400\,\text{cm}$, $D = 20\,\text{cm}$ and $M^2 = 36\,\text{cm}^2$ $N = 40$ and then $h = 10\,\text{cm}$. We can hardly distinguish the finite difference solution and the exact solution.

We find $k_\infty = 1.00155335$ in the approximate case and $k_\infty = 1.00155355$ in the exact case.

Then our approach is validated.

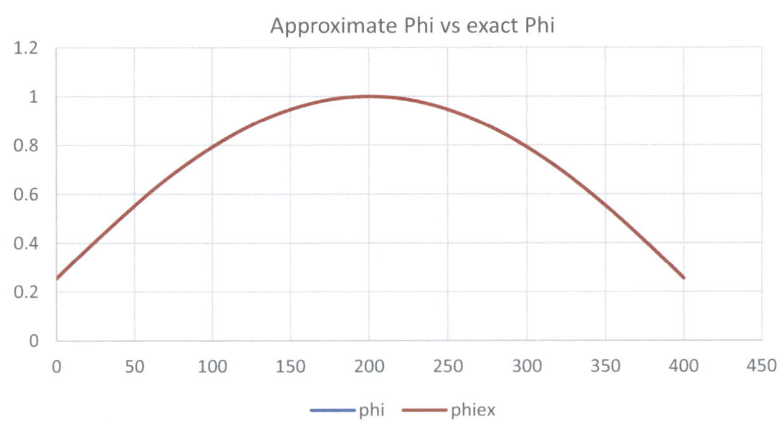

Fig. 17.2 Finite difference solution versus exact solution

17.2 Implementation of the Fourier BC with the 2 Groups Model

We come back to our reduced two groups model (10.13) and (10.14)

$$-M_1^2\frac{\partial^2}{\partial z^2}\psi+\left(1-k_{\infty,1}\right)\psi=k_{\infty,2}\Phi \tag{17.4}$$

$$-M_2^2\frac{\partial^2}{\partial z^2}\Phi+\Phi=\psi \tag{17.5}$$

We complement it with the Fourier boundary conditions

$$\psi(0)-2D_1\psi'(0)=0 \tag{17.6}$$

$$\psi(H)+2D_1\psi'(H)=0 \tag{17.7}$$

$$\Phi(0)-2D_2\Phi'(0)=0 \tag{17.8}$$

$$\Phi(H)+2D_2\Phi'(H)=0 \tag{17.9}$$

Like in the one group case, let $h=H/N$, and $(\psi_i)_{i=0,\dots,N}$, $(\Phi_i)_{i=0,\dots,N}$ be finite difference approximations of ψ and Φ, we introduce the matrices

$$A_1=\begin{pmatrix}\frac{2}{h^2}+\frac{1}{D_1h} & -\frac{2}{h^2} & & & \\ \cdots & \cdots & \cdots & \cdots & \cdots\\ & -\frac{1}{h^2} & \frac{2}{h^2} & -\frac{1}{h^2} & \\ \cdots & \cdots & \cdots & \cdots & \cdots\\ & & & -\frac{2}{h^2} & \frac{2}{h^2}+\frac{1}{D_1h}\end{pmatrix}$$

$$A_2=\begin{pmatrix}\frac{2}{h^2}+\frac{1}{D_2h} & -\frac{2}{h^2} & & & \\ \cdots & \cdots & \cdots & \cdots & \cdots\\ & -\frac{1}{h^2} & \frac{2}{h^2} & -\frac{1}{h^2} & \\ \cdots & \cdots & \cdots & \cdots & \cdots\\ & & & -\frac{2}{h^2} & \frac{2}{h^2}+\frac{1}{D_2h}\end{pmatrix}$$

$$B_1=M_1^2A_1+\left(1-k_{\infty,1}\right)I$$

$$B_2=M_2^2A_2+I$$

So that

$$B_1\psi = k_{\infty,2}\Phi$$

$$B_2\Phi = \psi$$

and

$$B_1 B_2 \Phi = k_{\infty,2}\Phi.$$

Then $k_{\infty,2}$ is an eigenvalue of matrix $B_1 B_2$.

Numerical Example

We select $k_{\infty,1} = 0.1$, $H = 400\,\mathrm{cm}$, $D_1 = 8.4\,\mathrm{cm}$, $D_2 = 1.6\,\mathrm{cm}$, $\Sigma_1 = 0.0175\,\mathrm{cm}^{-1}$ and $\Sigma_2 = 0.08\,\mathrm{cm}^{-1}$ so that

$$M_1^2 = 480\,\mathrm{cm}^2 \quad \text{and} \quad M_2^2 = 20\,\mathrm{cm}^2$$

We find $k_{\infty,2} = 0.92654$.

The neutron flux Φ in group 2 and the reduced neutron flux ψ in group 1 are represented on Fig. 17.3.

As predicted in Remark 10.2, we notice that they are almost equal except close to the boundary, where a boundary layer appears for the neutron flux Φ in group 2.

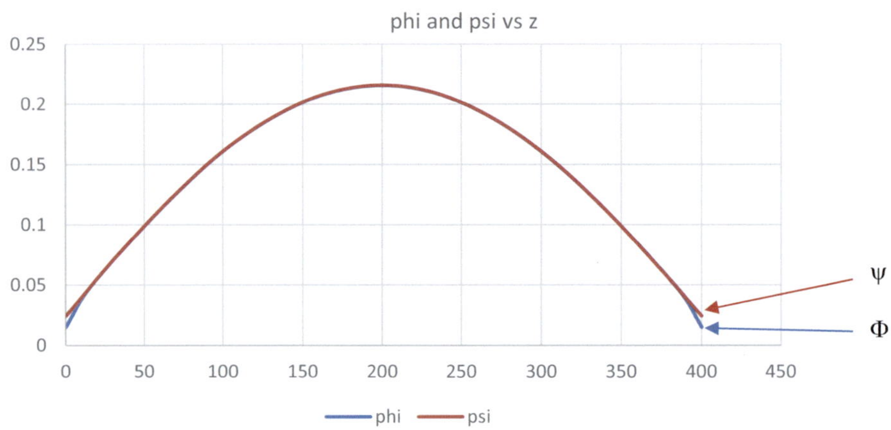

Fig. 17.3 Finite difference solution of the two groups model

17.3 Xenon Instabilities with the One Group Model [18]

Like in Exercise 7.3 the core is assumed to be homogeneous so that, in each cm^3 of core, there are ω cm^3 of fuel and $(1-\omega)$ cm^3 of moderator. In a PWR we have $\omega \sim 1/3$ so that the moderation ratio is $(1-\omega)/\omega \sim 2$.

Since there are X at/cm^3 of ${}^{135}_{54}Xe$ in the fuel and then ωX at/cm^3 in the core, the Xenon macroscopic cross section is equal to $\omega\sigma X$.

Let Σ denote the absorption cross section of the core without Xenon, we let $\alpha = \omega\sigma/\Sigma$. To take into account the neutron capture by ${}^{135}_{54}Xe$, the one group model writes:

$$-M^2\frac{\partial^2}{\partial z^2}\Phi + (1+\alpha X)\Phi = k_\infty\Phi, \quad 0 < z < H \tag{17.10}$$

When studying Xenon instabilities, we shall have $X = X(z,t)$ and then $\Phi = \Phi(z,t)$; more precisely $\Phi(z,t)$ will denote the neutron flux at elevation z and time t.

Since (17.10) is an eigenvalue problem, Φ is defined up to a multiplicative constant. We shall assume that

$$\Phi_{\text{ave}} = \frac{1}{H}\int_0^H \Phi(z,t)\mathrm{d}z \tag{17.11}$$

Is given for all t, which means that the core power is fixed. To (17.10) and (17.11) we add the following Bateman equations for Iodine and Xenon

$$\frac{\mathrm{d}Y}{\mathrm{d}t} = \gamma\Sigma_f\Phi - \lambda Y \tag{17.12}$$

$$\frac{\mathrm{d}X}{\mathrm{d}t} = \gamma'\Sigma_f\Phi + \lambda Y - (\mu + \sigma\Phi)X \tag{17.13}$$

where

- Y (resp. X) is the concentration of ${}^{135}_{53}I$ (resp. ${}^{135}_{54}Xe$) in the fuel evaluated in at/ cm^3 of fuel.
- σ is the microscopic capture cross section for Xenon in the thermal domain (evaluated in barns).
- Σ_f is the macroscopic fission cross section of the fuel in the thermal domain.
- γ the Iodine fission yield (about equal to 0.064).
- γ' the Xenon fission yield (about equal to 0.001).
- λ (resp. μ) the time constant for the β-decay of ${}^{135}_{53}I$ (resp. ${}^{135}_{54}Xe$).

For the sake of simplicity, in the following, we shall neglect Xenon generation from fission reaction directly, which is possible since $\gamma' \ll \gamma$.

We need to complement (17.10) with some boundary conditions in $z = 0$ and $z = H$. In the following we shall choose the Dirichlet BC:

$$\Phi(0) = \Phi(H) = 0 \tag{17.14}$$

We note that (17.12) and (17.13) are ordinary differential equations depending on parameter z.

We need to complement them with some initial conditions:

$$Y(z, 0) = Y_0(z) \tag{17.15}$$

$$X(z, 0) = X_0(z) \tag{17.16}$$

Rescaling of the One Group Model

Following [18] we let

$$\Phi(z, t) = \Phi_* \varphi(z, t)$$

$$X(z, t) = X_* x(z, t)$$

$$Y(z, t) = Y_* y(z, t)$$

We also replace t by $\tau = \lambda t$, and we let $\Phi_* = \lambda/\sigma$, $a = \mu/\lambda$ and $X_* = \frac{\gamma \Sigma_f \Phi_*}{\lambda} = \frac{\gamma \Sigma_f}{\sigma} = Y_*$.

We get the simpler system

$$\frac{dy}{d\tau} = \varphi - y \tag{17.17}$$

$$\frac{dx}{d\tau} = y - (a + \varphi)x \tag{17.18}$$

Finally, let $\alpha_* = \alpha X_* = \frac{\omega\sigma}{\Sigma} \cdot \frac{\gamma \Sigma_f}{\sigma} = \omega \cdot \frac{\gamma \Sigma_f}{\Sigma}$, we get

$$-M^2 \frac{\partial^2}{\partial z^2}\varphi + (1 + \alpha_* x)\varphi = k_\infty \varphi, \quad 0 < z < H \tag{17.19}$$

X_* has the following interpretation: it is the equilibrium value for the concentration of ^{135}Xe atoms when the neutron flux is infinitely large.

(17.17), (17.18) and (17.19) completed with Dirichlet BC and

Table 17.1 Typical numerical data for our one group model (see [9], p. 382)

λ	$2.92 \times 10^{-5}\,\mathrm{s}^{-1}$	Σ_f	$0.246\,\mathrm{cm}^{-1}$
μ	$2.12 \times 10^{-5}\,\mathrm{s}^{-1}$	Σ	$0.115\,\mathrm{cm}^{-1}$
a	0.725	Φ_*	$1.459 \times 10^{13}\,\mathrm{n/cm^2/s}$
γ	0.064	Y_*	$7.87 \times 10^{15}\,\mathrm{at/cm^3}$
σ	2×10^6 barns	X_*	$7.87 \times 10^{15}\,\mathrm{at/cm^3}$
ω	1/3	α_*	0.0452

$$\frac{1}{H}\int_0^H \varphi(z,t)\mathrm{d}z = \overline{\varphi}$$

where $\overline{\varphi}$ is given, is the one group system which we shall solve (Table 17.1).

We could have also changed the space variable $z \to Z/M$ to prove that the main parameter is H/M.

Space Approximation

We introduce a discretization step $\Delta z = H/(N+1)$ where N is the number of discretization points. We now denote by φ the column-vector of values of the rescaled thermal neutron flux at the discretization points $i\,\Delta z$, $1 \le i \le N$. (Our numerical results will be obtained with $N = 39$, but this is not a sensitive parameter.)

We introduce the tridiagonal matrix A such that

$$(A\Phi)_i = (-\,\Phi_{i-1} + 2\Phi_i - \Phi_{i+1})/\Delta z^2$$

where we assume that $\Phi_0 = 0$ and $\Phi_{N+1} = 0$ which corresponds to the Dirichlet boundary conditions.

(By using Sect. 17.1, it would be easy to shift from the Dirichlet BC to the Fourier BC.)

We shall follow the following convention: $(x\varphi)_i = x_i\varphi_i$, which means that, depending on the context x may denote either a vector in $\mathbb{R}^N$ or the $N \times N$ diagonal matrix containing the components of x on its main diagonal.

Let $\{y_\infty, x_\infty, \varphi_\infty\}$ be a steady solution of (17.17), (17.18) and (17.19) then satisfying

$$(a + \varphi_\infty)x_\infty = y_\infty = \varphi_\infty \tag{17.20}$$

$$M^2 A\varphi_\infty + (1 + \alpha_* x_\infty)\varphi_\infty = k_\infty\varphi_\infty \tag{17.21}$$

We note that (as above) the solution $\{y_\infty, x_\infty, \varphi_\infty\}$ of (17.20) and (17.21) cannot be unique: in fact x_∞ being given, (17.22) is an eigenvalue problem (which means that the core is critical) so that φ_∞ is defined up to a multiplicative constant and we fix this constant so that the core power is given.

In the case where our steady solution $\{y_\infty, x_\infty, \varphi_\infty\}$ is not stable, if we start from $\{y_a, x_a\}$ close to $\{y_\infty, x_\infty\}$ we shall activate an oscillation.

Time Discretization: Semi Implicit Scheme

We denote the time step by δ. We let $\tau_n = n\delta$.

We use the following semi implicit scheme:

First step, solve

$$y^n - \delta\big(\varphi^{n-1} - y^n\big) - y^{n-1} = 0$$

$$x^n - \delta\big(y^n - \big(a + \varphi^{n-1}\big)x^n\big) - x^{n-1} = 0$$

Second step, solve

$$\big(M^2 A + \big(I + \alpha_* x^n\big)\big)\varphi^n = k_\infty^n \varphi^n$$

Which produces both a value for k_∞^n (the smallest eigenvalue) and a (non-unique) value for φ^n.

Then we replace φ^n by φ^n times $\overline{\varphi}/\text{average}(\varphi^n)$, where $\overline{\varphi}$ is a prescribed value for the average of the (normalized) flux.

Results with the One Group Model

When the average flux is sufficiently high the flux oscillations are growing.

A typical result is given in Fig. 17.4 for which $\overline{\varphi} = 3$ which is a rather high value since $3\Phi_* = 4.38 \times 10^{13}$ n/cm^2/s (note that Fig. 17.4 shows oscillations).

To illustrate the evolution of the oscillations we shall use the Axial Offset AO_p defined by $\text{AO}_\text{p} = (P_\text{t} - P_\text{b})/(P_\text{t} + P_\text{b})$ where P_t (resp. P_b) is the average power (or neutron thermal flux) in the upper half part of the core (resp. lower half part).

Figure 17.4 shows the evolution of AO_p with the reduced time τ.

We see that the oscillations are

- damped out when $M^2 = 100\,\text{cm}^2$ and $\overline{\varphi} = 1$
- stable when $M^2 = 100\,\text{cm}^2$ and $\overline{\varphi} = 2$ or $M^2 = 64\,\text{cm}^2$ and $\overline{\varphi} = 1$
- growing when $M^2 = 100\,\text{cm}^2$ and $\overline{\varphi} = 3$ or $M^2 = 64\,\text{cm}^2$ and $\overline{\varphi} =$ 2 or 3(Fig. 17.5).

On a standard PWR we know that the migration area is more like $64\,\text{cm}^2$. Note that the period of oscillations depends both on the migration area and on $\overline{\varphi}$: it can range between 28 and 36 h, which is consistent with the measurements made in actual PWR (~ 30 h for a 900 MWe [19]).

When the height of the reactor core is less than 250 cm, no oscillation is activated (see Fig. 17.6). Once the height is more than 300 cm, a growing oscillation can be produced. Moreover, the higher the height is, the bigger AO_p is.

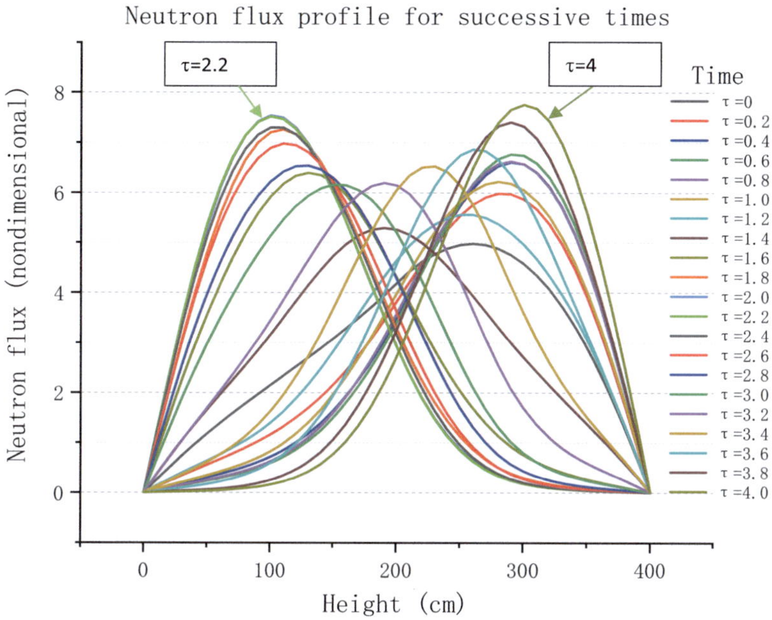

Fig. 17.4 The oscillation of neutron flux in 1 group model with $M^2 = 64\ \text{cm}^2$

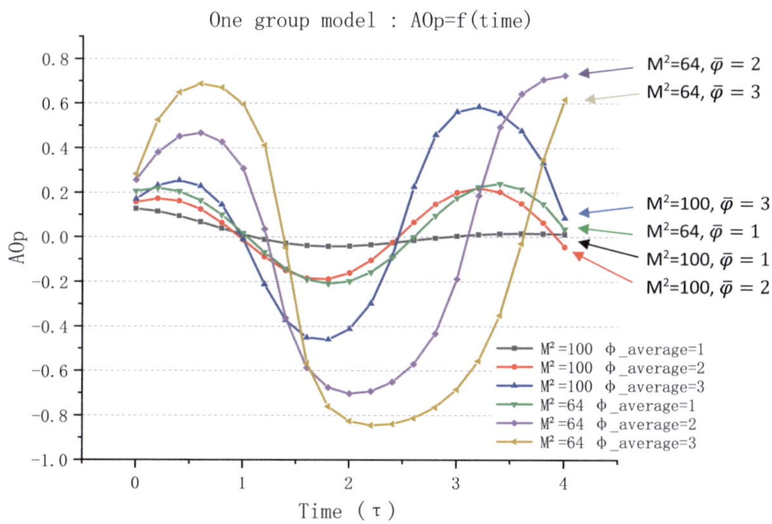

Fig. 17.5 Evolution of AO_p for various values of $\overline{\varphi}$ and M^2

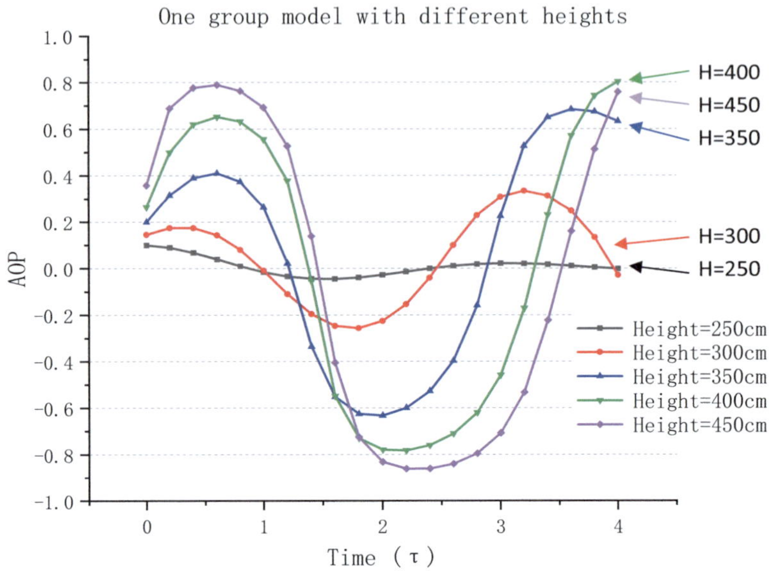

Fig. 17.6 Evolution of AO_p for various values of height

17.4 Xenon Instabilities with the 2 Groups Model [18]

The core is still assumed to be homogeneous so that, in each cm^3 of core, there are ω cm^3 of fuel and $(1-\omega)$ cm^3 of moderator. In a PWR we have $\omega \sim 1/3$ so that the moderation ratio $(1-\omega)/\omega \sim 2$.

The fact is that the Xenon capture cross section is negligible for fast neutrons ($E > 10$ eV).

If we select in an adequate way the limit between fast and thermal neutrons we shall be able to use the system of coupled diffusion equations that was the subject of Exercise 10.1

$$-D_1 \frac{\partial^2}{\partial z^2}\Psi + \Sigma_1 \Psi = \nu\omega\Sigma_{f2}\Phi + \nu\omega\Sigma_{f1}\Psi \tag{17.22}$$

$$-D_2 \frac{\partial^2}{\partial z^2}\Phi + \Sigma_2 \Phi + \omega\sigma X \Phi = p\Sigma_1 \Psi \tag{17.23}$$

where X is the number of ${}^{135}_{54}Xe$ atoms per fuel cm^3. (Note that in Chap. 10 it was per core cm^3.)

Like in Sect. 17.3, $\omega\sigma X$ is the xenon macroscopic capture cross section: this is why it is added to Σ_2.

In the same way, if Σ_{f1} (resp. Σ_{f2}) denotes the macroscopic fission cross section of the fuel in the fast (resp. thermal) neutron domain, $\omega\Sigma_{f1}\Psi$ (resp. $\omega\Sigma_{f2}\Phi$) denotes the number of fast (resp. thermal) fission per cm^3 per s in the core.

Now, as in Sect. 10, $\Sigma_1\Psi$ is the number of fast neutrons which disappear per cm^3 per s. Some of them are physically absorbed but a fraction p of them reappear, after slowing down, in the thermal neutron group.

The factor p is then the resonance escape probability from Fermi's four factors formula.

For the sake of simplicity, we shall assume that the diffusion coefficients do not depend on the space variable z.

Even though (17.22) and (17.23) are stationary diffusion equations, as in Sect. 17.3, we have $X = X(z, t)$.

The xenon evolution is coupled to the iodine and Xenon evolution through

$$\frac{\mathrm{d}Y}{\mathrm{d}t} = \gamma(\Sigma_{f1}\Psi + \Sigma_{f2}\Phi) - \lambda Y \tag{17.24}$$

$$\frac{\mathrm{d}X}{\mathrm{d}t} = \lambda Y - (\mu + \sigma\Phi)X \tag{17.25}$$

Such an evolution is rather slow at the scale of one hour, so there is no need to consider the neutron evolution at the timescale of delayed neutrons.

The 2 groups model will be complemented by some boundary conditions for the neutron flux.

For simplicity, we shall consider the Dirichlet boundary conditions

$$\Psi(0) = \Psi(H) = \Phi(0) = \Phi(H) \tag{17.26}$$

but it is also possible to consider the Fourier boundary conditions as in Sect. 17.2.

Remark 17.1 The number of neutrons produced by fast fission of $^{238}_{92}$U is actually 2.8, but, for the sake of simplicity, we choose $\nu = 2.4$ both for $^{238}_{92}$U and $^{235}_{92}$U. However, it would be easy to change the term $\nu\omega\Sigma_{f1}\Psi$ in (17.22) by $\nu_1\omega\Sigma_{f1}\Psi$ where ν_1 is some appropriate average of 2.4 and 2.8 …

Rescaling of the 2 Groups Model

Like in Sect. 17.3 the system (17.22)–(17.25) can be simplified.

At first, let $M_1^2 = \frac{D_1}{\Sigma_1}$, $M_2^2 = \frac{D_2}{\Sigma_2}$, $k_{\infty,1} = \frac{\nu\omega\Sigma_{f1}}{\Sigma_1}$, $k_{\infty,2} = \frac{\nu\omega p\Sigma_{f2}}{\Sigma_2}$ and $\alpha = \frac{\omega\sigma}{\Sigma_2}$.

Then, we rewrite the variables $\Phi = \Phi_*\varphi$, $\Psi = \Psi_*\psi$, $X = X_*x$, $Y = Y_*y$, $\tau = \lambda t$.

And we obtain

$$-\Psi_* M_1^2 \frac{\partial^2}{\partial z^2}\Psi + \Psi_*\psi = k_{\infty,1}\Psi_*\psi + k_{\infty,2}\frac{\Sigma_2}{p\Sigma_1}\Phi_*\varphi$$

$$
-\Phi_* M_2^2 \frac{\partial^2}{\partial z^2}\varphi + \Phi_*\varphi + \alpha X_* \Phi_* x\varphi = \frac{p\Sigma_1}{\Sigma_2}\Psi_*\psi
$$

$$
\lambda Y_* \frac{\mathrm{d}y}{\mathrm{d}\tau} = \gamma(\Sigma_{\mathrm{f1}}\Psi_*\psi + \Sigma_{\mathrm{f2}}\Phi_*\varphi) - \lambda Y_* y
$$

$$
\lambda X_* \frac{\mathrm{d}x}{\mathrm{d}\tau} = \lambda Y_* y - (\mu + \sigma\Phi_*\varphi)X_* x
$$

To simplify the above equations, we choose $X_* = Y_*$ and $\Phi_* = \frac{p\Sigma_1}{\Sigma_2}\Psi_*$

$$
-M_1^2 \frac{\partial^2}{\partial z^2}\psi + \psi = k_{\infty,1}\psi + k_{\infty,2}\varphi
$$

$$
-M_2^2 \frac{\partial^2}{\partial z^2}\varphi + (1 + \alpha X_* x)\varphi = \psi
$$

$$
\lambda Y_* \frac{\mathrm{d}y}{\mathrm{d}\tau} = \gamma(\Sigma_{\mathrm{f1}}\Psi_*\psi + \Sigma_{\mathrm{f2}}\Phi_*\varphi) - \lambda Y_* y
$$

$$
\lambda \frac{\mathrm{d}x}{\mathrm{d}\tau} = \lambda y - (\mu + \sigma\Phi_*\varphi)x
$$

If we choose

$$
\Phi_* = \frac{\lambda}{\sigma},\ a = \frac{\mu}{\lambda},\ X_* = \frac{\gamma\Sigma_{\mathrm{f2}}\Phi_*}{\lambda} = \frac{\gamma\Sigma_{\mathrm{f2}}}{\sigma},
$$

$$
\zeta = \frac{\gamma\Sigma_{\mathrm{f1}}\Psi_*}{\lambda X_*} = \frac{\Sigma_{\mathrm{f1}}\Sigma_2}{p\Sigma_1\Sigma_{\mathrm{f2}}} = \frac{k_{\infty,1}}{k_{\infty,2}},\ \alpha_* = \alpha X_* = \frac{\omega\sigma}{\Sigma_2}\frac{\gamma\Sigma_{\mathrm{f2}}}{\sigma} = \frac{\gamma\omega\Sigma_{\mathrm{f2}}}{\Sigma_2}
$$

We can simplify the system one more time and we have

$$
-M_1^2 \frac{\partial^2}{\partial z^2}\psi + \psi = k_{\infty,1}\psi + k_{\infty,2}\varphi = k_{\infty,2}(\zeta\psi + \varphi) \tag{17.27}
$$

$$
-M_2^2 \frac{\partial^2}{\partial z^2}\varphi + (1 + \alpha_* x)\varphi = \psi \tag{17.28}
$$

$$
\frac{\mathrm{d}y}{\mathrm{d}\tau} = \zeta\psi + \varphi - y \tag{17.29}
$$

$$
\frac{\mathrm{d}x}{\mathrm{d}\tau} = y - (a + \varphi)x \tag{17.30}
$$

Space Approximation

We introduce the tridiagonal matrix A as in Sect. 17.3.

Let $\{y_\infty, x_\infty, \varphi_\infty, \psi_\infty\}$ be a steady solution of (17.27), (17.28), (17.29) and (17.30) satisfying

$$(a + \varphi_\infty)x_\infty = y_\infty = \zeta\psi_\infty + \varphi_\infty \tag{17.31}$$

$$M_1^2 A\psi_\infty + (1 - k_{\infty,1})\psi_\infty - k_{\infty,2}\varphi_\infty = 0 \tag{17.32}$$

$$M_2^2 A\varphi_\infty + (1 + \alpha_* x_\infty)\varphi_\infty - \psi_\infty = 0 \tag{17.33}$$

After elimination of ψ_∞ thanks to (17.33) we find that, $k_{\infty,1}$ and x_∞ being given, $k_{\infty,2}$ is the smallest eigenvalue of

$$\big(M_1^2 A + (1 - k_{\infty,1})I\big)\big(M_2^2 A + (I + \alpha_* x_\infty)\big)\varphi_\infty = k_{\infty,2}\varphi_\infty$$

Therefore φ_∞ and ψ_∞ are not unique: they are defined up to a scalar multiplicative factor that can be fixed, given the core power.

In the case where our steady solution $\{y_\infty, x_\infty, \varphi_\infty, \psi_\infty\}$ is not stable, if we start from $\{y_a, x_a\}$ close to $\{y_\infty, x_\infty\}$ we shall activate an oscillation.

To simulate such a perturbation, the xenon concentration will be increased by 10% in the half lower part of the core. In the half upper part, it will be decreased by 10%.

Time Discretization: Semi Implicit Scheme

We proceed as in Sect. 17.3 and introduce the following semi implicit scheme:

First step, solve

$$y^n - \delta\big(\zeta\psi^{n-1} + \varphi^{n-1} - y^n\big) - y^{n-1} = 0$$

$$x^n - \delta\big(y^n - \big(a + \varphi^{n-1}\big)\big)x^n - x^{n-1} = 0$$

Second step, solve

$$\big(M_1^2 A + (1 - k_{\infty,1})I\big)\big(M_2^2 A + \big(I + \alpha_* x^n\big)\big)\varphi^n = k_{\infty,2}\varphi^n$$

Which produces both a value for $k_{\infty,2}^n$ and a (non-unique) value for φ^n.

Then we replace φ^n by φ^n times $\overline{\varphi}/\text{average}(\varphi^n)$, where $\overline{\varphi}$ is a prescribed value for the average of the (normalized) flux.

The set of values given in Table 17.2 is compatible for 3.7% enriched UO_2 with results given by TRIPOLI4 and the 6 factors formula established in Santamarina [20] that is

$$k_\infty = \chi \cdot \varepsilon_{\text{even}} \cdot \varepsilon_{\text{odd}} \cdot p \cdot f \cdot \eta.$$

Table 17.2 Typical numerical data for 2 groups model

Σ_{f1}	$0.007\,\text{cm}^{-1}$	$k_{\infty,2}$	1.017
Σ_{f2}	$0.246\,\text{cm}^{-1}$	α_*	0.0452
Σ_1	$0.025\,\text{cm}^{-1}$	ζ	0.274
Σ_2	$0.119\,\text{cm}^{-1}$	a	0.725
ν	2.4	Ψ_*	$9.870 \times 10^{13}\,\text{n/cm}^2/\text{s}$
p	0.616	Φ_*	$1.459 \times 10^{13}\,\text{n/cm}^2/\text{s}$
$k_{\infty,1}$	0.224		

Compared with Fermi's 4 factors formula, this formula replaces ε by the product $\chi \cdot \varepsilon_{\text{even}} \cdot \varepsilon_{\text{odd}}$ where χ takes into account the $(n, 2n)$ reactions, $\varepsilon_{\text{even}}$ the $^{238}_{92}\text{U}$ fissions, and ε_{odd} the fast fissions of $^{235}_{92}\text{U}$. With such a definition, we easily check that $k_{\infty,2} = p \cdot f \cdot \eta$.

Numerical Results

The power of the reactor will be maintained constant which means the average neutron flux is fixed. $\varphi_{\text{ave}} = \overline{\varphi} = 3$ so that $\Phi_{*,\text{ave}} = 3\Phi_* = 4.38 \times 10^{13}\,\text{n/cm}^2/\text{s}$. (This is somewhat too high for a PWR reactor.)

In Fig. 17.7 we give the thermal flux profile as a function of time. This should be compared to Fig. 17.4. We notice that when $M^2 = M_1^2 + M_2^2$ (which is the case here) the results given by the one group and the two groups models look pretty much the same.

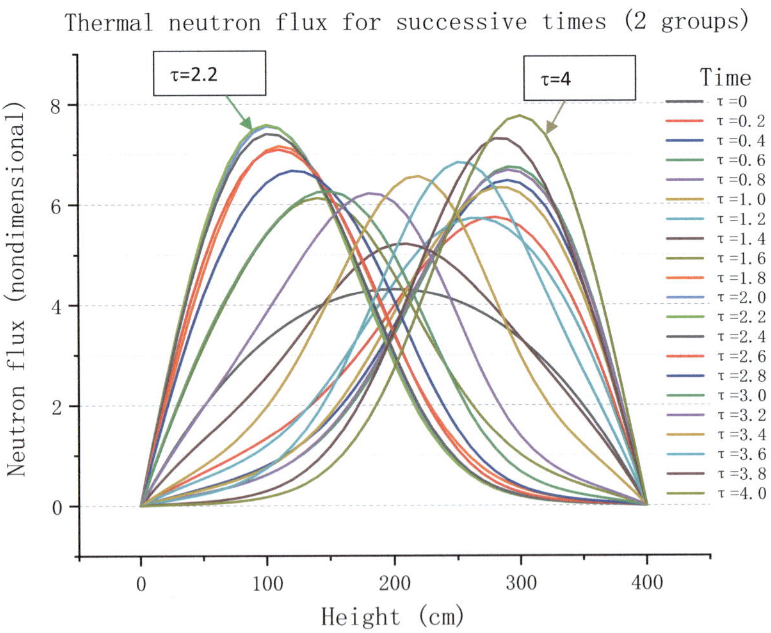

Fig. 17.7 The oscillation of neutron flux in 2 groups model with $M_1^2 = 55\,\text{cm}^2$ $M_2^2 = 9\,\text{cm}^2$

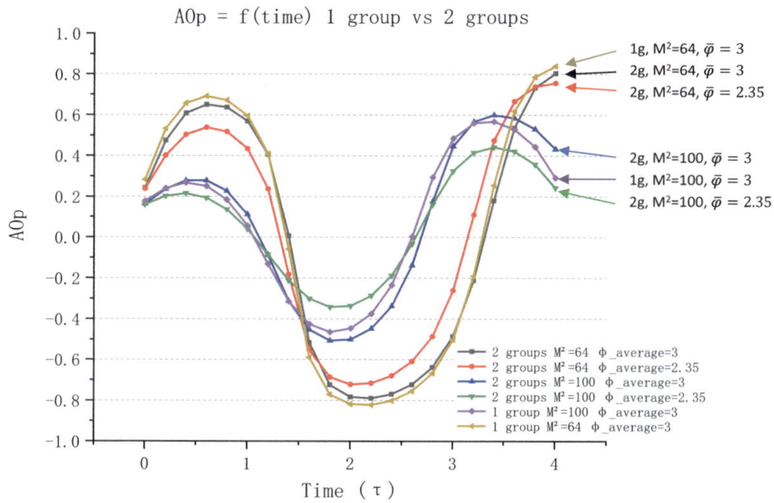

Fig. 17.8 One group versus two groups

(We have observed in our results that $\varphi \cong \psi$, but this is no surprise when we consider (17.28): this results from $\alpha_* x \ll 1$ and from $M_2 \ll H$.)

On Fig. 17.8 we compare one group and two group results. We plot AO_p as a function of the reduced time τ.

We see that the behavior of one and two groups models is pretty much the same.

We check again that the migration area has a significant influence on the time-period of oscillations.

17.5 The Nonlinear Model (Power Feedback Effect)

Most water moderated reactors have a negative temperature coefficient, which means that when the core temperature increases the reactivity of the core decreases.

As we have seen in Chaps. 15 and 16, for lightly enriched Uranium fuel, there is a negative fuel temperature coefficient due to the Doppler effect.

Controlling the chain reaction is obviously of primary importance for nuclear safety.

Such negative temperature effects are then very useful.

Like in [18] we shall use the following nonlinear model:

$$-M^2 \frac{\partial^2}{\partial z^2}\varphi + (1 + \alpha_* x)\varphi = k_\infty (1 - \beta\varphi)\varphi \tag{17.34}$$

$$\frac{\mathrm{d}y}{\mathrm{d}\tau} = \varphi - y \tag{17.35}$$

$$\frac{dx}{d\tau} = y - (a + \varphi)x \tag{17.36}$$

β is the prompt power feedback coefficient.

Note that, unlike what we had in the linear case (see 17.10), φ is uniquely defined by (17.34) and the BC.

The nonlinear Eq. (17.34) describes what happens with the fuel (Doppler) temperature effect.

To take into account the moderator temperature effect, we should compute the moderator temperature locally, which would mean coupling neutron diffusion equations and thermal hydraulics.

To keep simple models, we shall not do that and we refer the reader to [21] for a simplified approach.

Space Approximation

Proceeding like in Sect. 17.3, we obtain the following system

$$\frac{dy}{d\tau} = \varphi - y \tag{17.37}$$

$$\frac{dx}{d\tau} = y - (a + \varphi)x \tag{17.38}$$

$$A\varphi + (1 + \alpha_* x)\varphi = k_\infty(1 - \beta\,\varphi)\varphi \tag{17.39}$$

Algorithm: We notice that (17.37)–(17.39) is a particular case of the following differential system:

$$\dot{y} = E(y, x, \varphi) \tag{17.40}$$

$$\dot{x} = F(y, x, \varphi) \tag{17.41}$$

$$0 = G(y, x, \varphi) \tag{17.42}$$

where $y(t), x(t), \varphi(t) \in \mathbb{R}^N$

$$E : \mathbb{R}^N \times \mathbb{R}^N \times \mathbb{R}^N \to \mathbb{R}^N$$

$$F : \mathbb{R}^N \times \mathbb{R}^N \times \mathbb{R}^N \to \mathbb{R}^N$$

$$G : \mathbb{R}^N \times \mathbb{R}^N \times \mathbb{R}^N \to \mathbb{R}^N$$

Which we should complement with initial conditions:

$$y(0) = y_a; \quad x(0) = x_a.$$

When we do not take the temperature effect into account ($i - e$ when $\beta = 0$) the RHS of (4.6) is just equal to $k_\infty \Phi$ which means that (17.39) is an eigenvalue problem. Therefore, it has a solution $\varphi \geq 0$ only if k_∞ is equal to the smallest eigenvalue of matrix

$$C = C(x) = A + (1 + \alpha_* x)$$

In the nonlinear case ($\beta > 0$) we can hope (and that's what we see in our simulations) that if k_∞ is larger that this smallest eigenvalue, (4.6) will have a solution φ, and that this solution is larger if k_∞ increases.

To identify (17.37)–(17.39) and (17.40)–(17.42) we need to choose

$$E(y, x, \varphi) = \varphi - y$$

$$F(y, x, \varphi) = y - (a + \varphi)x$$

$$G(y, x, \varphi) = A\varphi + (1 + \alpha_* x)\varphi - k_\infty(1 - \beta\varphi)\varphi.$$

Time Discretization: Fully Implicit Scheme

We have successfully used the following fully implicit scheme.

$$\left(y^n - y^{n-1}\right)/\delta = E\left(y^n, x^n, \varphi^n\right) \tag{17.43}$$

$$\left(x^n - x^{n-1}\right)/\delta = F\left(y^n, x^n, \varphi^n\right) \tag{17.44}$$

$$0 = G\left(y^n, x^n, \varphi^n\right) \tag{17.45}$$

This seems quite obvious, however we note that in a nuclear reactor, when the power is given, it means the average value of the flux is equal to $\overline{\varphi}$, the reactivity is adjusted to zero not by the Doppler effect, but by the boron concentration, which means that the control system selects k_∞^n as the smallest eigenvalue of matrix.

$$A + \left(1 + \alpha_* x^n\right) + k_\infty^{n-1}\beta\varphi^* \tag{17.46}$$

where $\varphi^* = \mu_n \varphi^n$ and $\mu_n = \overline{\varphi}/\text{average}(\varphi^n)$.

We leave to the reader the details of implementation of Newton's method to solve the non-linear system (17.43)–(17.45) of 3 N equations for 3 N unknowns.

Time Discretization: Semi Implicit Scheme

Step 1: Like in the fully implicit scheme, we evaluate the smallest eigenvalue k_∞^n of the matrix A defined in (17.46).

Step 2: Knowing y^n, x^n we compute φ^n by solving the $N \times N$ non-linear system

$$G(y^n, x^n, \varphi^n) = 0$$

(For this, we use Newton's method.)

Step 3: We compute y^{n+1}, x^{n+1} with:

$$(y^{n+1} - y^n)/\delta = E(y^{n+1}, x^{n+1}, \varphi^n)$$

$$(x^{n+1} - x^n)/\delta = F(y^{n+1}, x^{n+1}, \varphi^n)$$

Remark 17.2

(1) the algorithm we use in the linear case ($\beta = 0$) is basically the same, except that we avoid step 2.
(2) the results we obtained with both algorithms are almost the same.

How to Choose β?

In PWR 1300 MWe, the rated reactor power corresponds to $\overline{\varphi} \sim 3$.

The "power defect" in the EDF terminology is of the order of 1500 pcm. But we're only going to take the Doppler effect which is of the order of 1000 pcm. (In fact, differing from the Doppler effect, the moderator effect is not a local effect because of the water circulation in the core, and therefore an increase in its temperature certainly reduces the reactivity of the core, but over much of its height.)

This means that we consider $\beta \Phi_{\text{moy}} \cong 0.01$ (Fig. 17.9).

From Fig. 17.10, we shall see that, with the Doppler effect, the amplitude of the oscillation decreases.

Shimazu's Representation

Like Shimazu [22] we introduce another axial offset for Iodine

$$\text{AO}_Y = (Y_t - Y_b)/(Y_t + Y_b)$$

And a different one for Xenon

$$\text{AO}_X = (X_t - X_b)/(X_t + X_b - 2. * X_t * X_b)$$

NB. Note that at constant power the normalized Xenon level should be < 1 so that we have both $X_t < 1$ *and* $X_b < 1$, which proves that

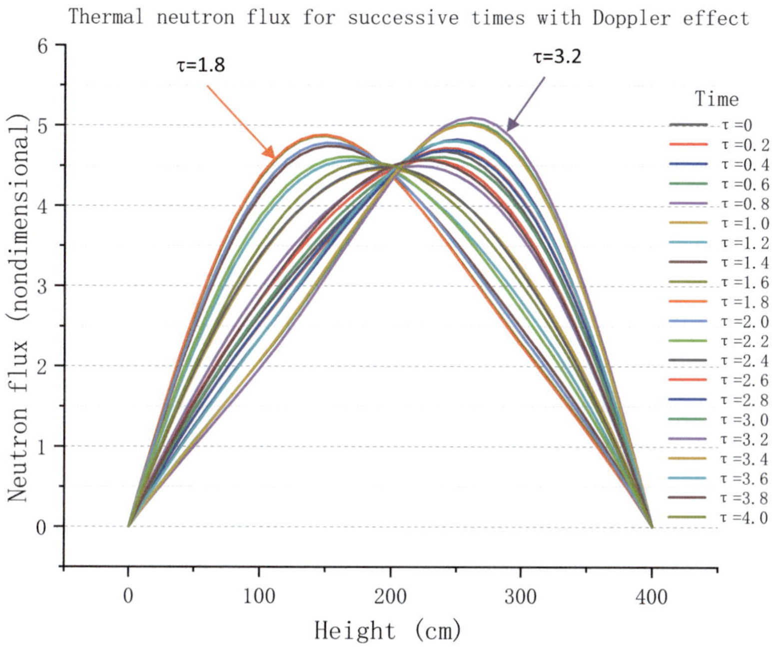

Fig. 17.9 The oscillation of neutron flux in 1 group model with the Doppler effect $\beta = 0.01/3$

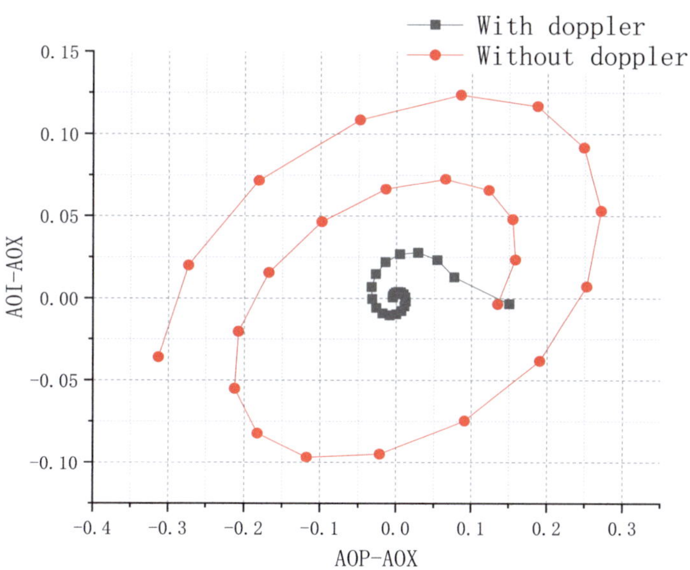

Fig. 17.10 Shimazu's representation with/without Doppler effect

$$X_t + X_b - 2. * X_t * X_b = X_t(1 - X_b) + X_b(1 - X_t) > 0$$

Shimazu's ellipses are a special representation of the results where we plot $AO_Y - AO_X$ with respect to $AO_p - AO_X$. Here is what we get with $M^2 = 100\,\text{cm}^2$, $\overline{\varphi} = 3$ and $\alpha_* = 0.0386$. The initial point starts from the horizontal axis. It turns in the trigonometric sense. Note that the curve is more a spiral than an ellipse.

The ellipse is divergent in case there is no Doppler effect, convergent with Doppler effect.

Comparing with the diverging curve of the model without the Doppler effect, we can draw a conclusion: the Doppler effect effectively weakens the oscillation. Actually, as we have seen in Chap. 15, the Doppler effect plays an important role in the inherent safety of reactor.

However, even if the oscillations are more limited than without Doppler effect, they still need to be controlled, and that's what we study in the next section.

17.6 Control Laws [18]

Shimazu's Control Method [22]

Shimazu has been the first to use the axial offsets AO_p, AO_x and AO_y to control the xenon instability.

When we drop the control rods, the representative point will move to the left because AO_p will decrease, however, if you raise the control bars, it will move right.

Note that we use lightly absorbing control rods, which means that when they are fully inserted, they reduce the reactivity of the core by 1100 pcm.

As in [22], we select 10 cm for a step.

Here are the criterions:

- If $AO_y - AO_x > 0$ and $\frac{AO_y - AO_x}{AO_p - AO_x} < \tan(\varphi_R)$ we raise the control rods for a step

  To improve the result, if $AO_y - AO_x > 0.2$ we raise the control rods for two steps

- If $AO_y - AO_x < 0$ and $\frac{AO_y - AO_x}{AO_p - AO_x} < \tan(\varphi_R)$ we drop the control rods for a step

  To improve the result, if $AO_y - AO_x < -0.2$ we drop the control rods for two steps.

To calculate φ_R, we should use the following formula:

$$\varphi_R = \frac{1}{2}\arctan\left[\frac{2r\sin(\varphi_p)}{1 - r^2}\right] r = \frac{AMC}{AM_p + AM_x}$$

$$AMC = \left(\left[AM_y \cdot \cos(\varphi_i) + AM_x \cdot \cos(\varphi_p)\right]^2 + \left[AM_y \cdot \sin(\varphi_i) + AM_x \cdot \sin(\varphi_p)\right]^2\right)^{1/2}$$

$$AO_x - AO_p = (AM_p + AM_x)e^{-\gamma t \cos(\omega t - \varphi_p)}$$

$$\mathrm{AO}_y - \mathrm{AO}_x = \mathrm{AMC}\,\mathrm{e}^{-\gamma t \sin(\omega t)}$$

φ_i and φ_p	phase at the initial time of AO_y and AO_p respectively
AM_p, AM_y, AM_x	amplitude of AO_p, AO_y and AO_x respectively
γ	stability index of the xenon oscillation
ω	angular velocity of oscillation.

At the initial moment, it has an oscillation because there is the disturbance and we used an application named Origin (https://www.originlab.com/) to get the sinusoidal functions of AOP and AOI, then we know φ_i and φ_p and finally φ_R.

On Fig. 17.11, we give what we get with Shimazu's method when $M^2 = 100\,\mathrm{cm}^2$. We can see that Shimazu's method eliminates the xenon oscillation quite well.

PID Method

We shall test a more standard PID regulator.

The principle is quite simple, it can be described as an equation.

$$U(t) = K_\mathrm{p}\left[\mathrm{err}(t) + \frac{1}{T_\mathrm{i}}\int \mathrm{err}(t)\mathrm{d}t + \frac{T_\mathrm{d} \times d\ \mathrm{err}(t)}{\mathrm{d}t}\right]$$

After time discretization, we can obtain a new formula

$$U(k) = K_\mathrm{p}e(k) + \frac{K_\mathrm{p}T}{T_\mathrm{i}}\sum_{n=0}^{k} e(n) + \frac{K_\mathrm{p}T_\mathrm{d}}{T}[e(k) - e(k-1)]$$

where

U	variable that can be controlled
err (or e)	error
T	control period
K_p	proportional coefficient.

In the following, we let $K_\mathrm{i} = \frac{K_\mathrm{p}T}{T_\mathrm{i}}$ and $K_\mathrm{d} = \frac{\mathrm{K_pT_d}}{\mathrm{T}}$.

In our case, U denotes the control rod's position, e denotes the axial offset AO_p, so we can describe the position of control bars by this equation:

$$\overline{x}_1(k) = 10\left(30 + K_\mathrm{p}\cdot \mathrm{AO}_\mathrm{p} + K_\mathrm{i}\sum_{n=0}^{k} e(n) + K_\mathrm{d}\cdot[e(k) - e(k-1)]\right)$$

Note that we have selected $\overline{x}_1 = 300$ as the initial position of our control bars.

On Fig 17.12 is the result with our PID method when $\mathrm{M2} = 100\,\mathrm{cm}^2$; $K_\mathrm{p} = -75$, $T_\mathrm{i} = 50$, $T_\mathrm{d} = 0.01$.

When we use PID regulator to control the system, we found that T_d is a sensitive parameter. Let's just consider the differential item: $K_\mathrm{d}\cdot[e(k) - e(k-1)]$.

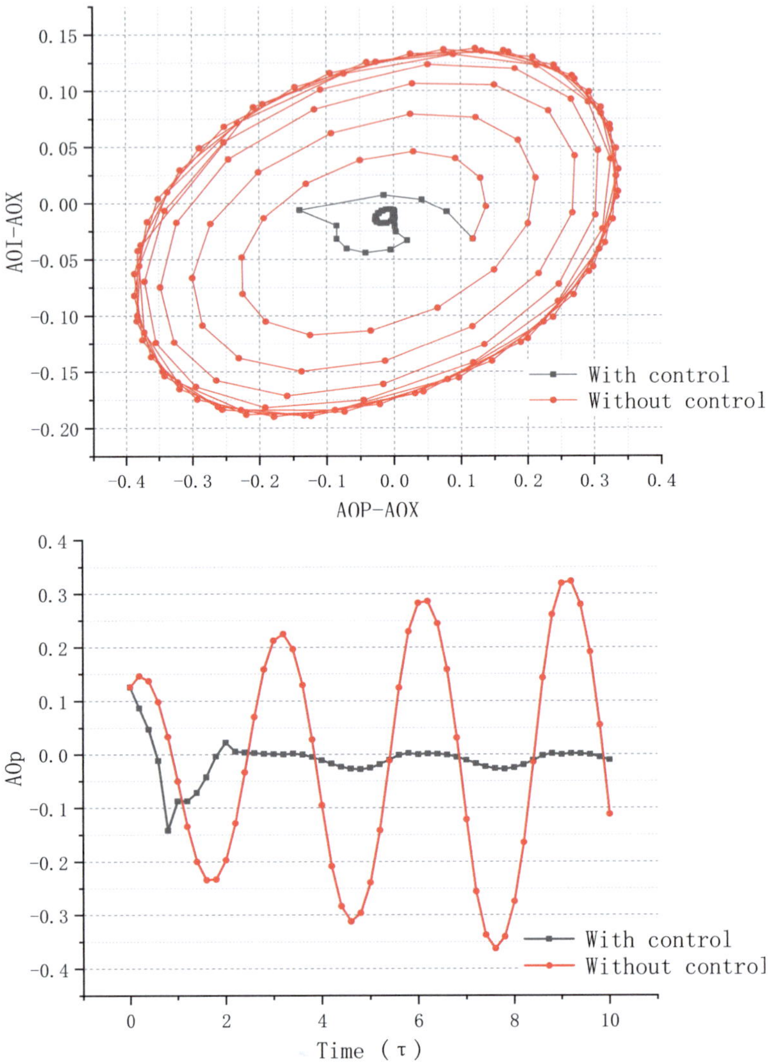

Fig. 17.11 Effect of Shimazu's control method

We let $T_i = 0$ and we neglect the proportional item, then we can describe the position of control bars by this equation: $\overline{x}_1(k) = \overline{x}_1(k-1) + 10K_d \cdot [e(k) - e(k-1)]$.

Which means, practically, that when AO_p increases, the control bars should be inserted.

We can see on Figs. 17.12 and 17.13 that an adequate PID regulator can also control the system quite well.

We note that K_d is the more sensitive parameter.

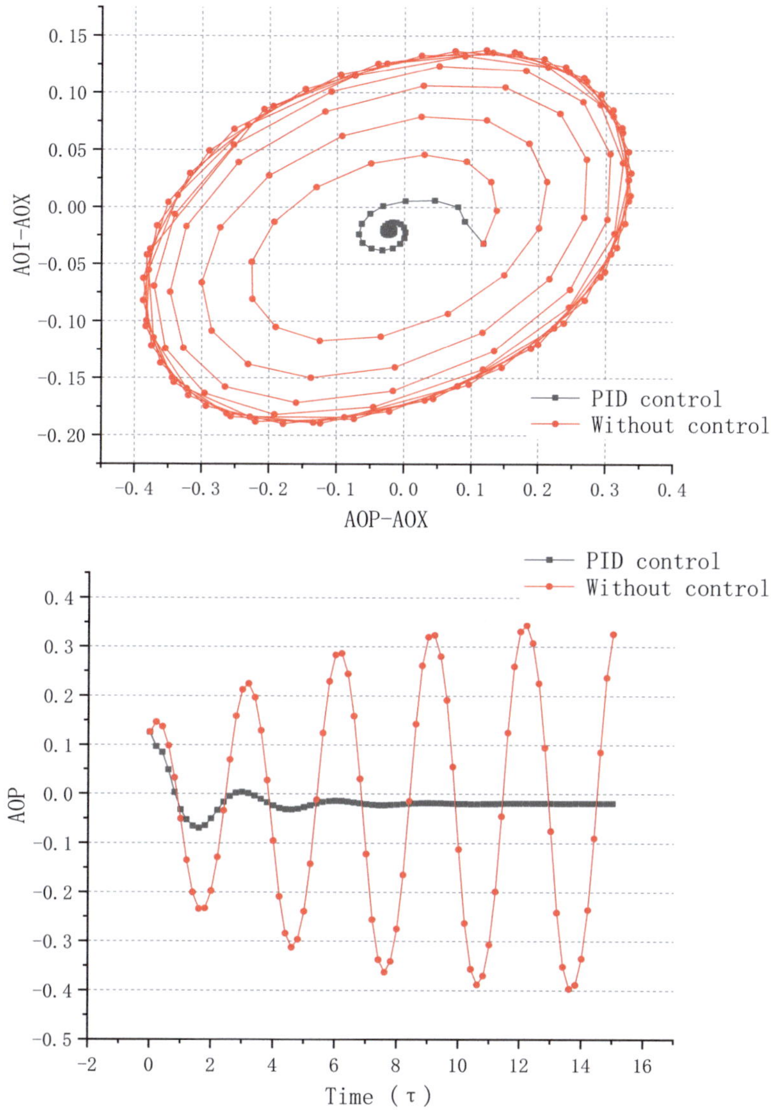

Fig. 17.12 Effect of our PID control strategy with $K_p = -75$, $T_i = 50$, $T_d = 0.01$

In conclusion to this chapter, even though Shimazu's control law has been designed by using a simple 2 nodes lumped model of the core, it works well also when applied to a more detailed one group finite difference model of the core.

In Sect. 17.4, we have studied a two groups model and show that it gives basically the same trend as the one group model, but the choice of physical parameters is more obvious with the two groups model.

Both models show that xenon oscillations are significant on large PWR cores.

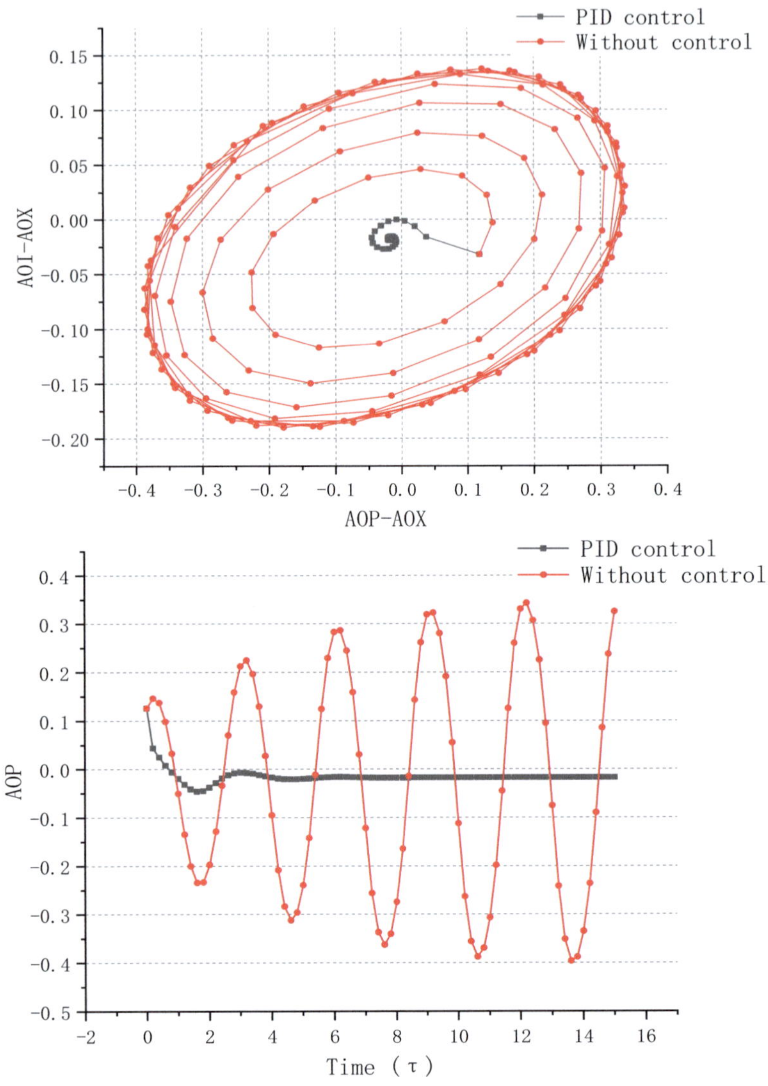

Fig. 17.13 Effect of our PID control strategy with $K_d = 130$

The oscillations are reduced when some nonlinear effects like the Doppler effect are taken into account (see Sect. 17.5).

However, the xenon oscillations still need to be controlled.

In Sect. 17.6, we show that Shimazu's method is efficient, but more standard methods like the widely used PID regulator are also efficient.

Annex 1: Proof of Formula (6.1)

Calculations are simpler if we pass through the center of mass, which moves at the speed of

$$\vec{v}_G = \frac{m_A}{m_B + m_A}\vec{v}$$

In fact, the velocity of the center of gravity of the two particles is identical after impact.

Particle velocities of A and B in the center-of-mass reference frame:

$$\vec{v}_A = \vec{v} - \vec{v}_G = \frac{m_B}{m_B + m_A}\vec{v}$$
$$\vec{v}_B = \vec{0} - \vec{v}_G = -\frac{m_A}{m_B + m_A}\vec{v}$$

Let $\vec{v}'_A$ (resp. $\vec{v}'_B$) denote the velocity of A (resp. B) after collision in the center of mass frame.

Conservation of momentum:

$$m_A\vec{v}_A + m_B\vec{v}_B = m_A\vec{v}'_A + m_B\vec{v}'_B$$

i.e.,

$$m_A\vec{v}_A + m_B\vec{v}_B = \frac{m_A m_B}{m_B + m_A}\vec{v} - \frac{m_B m_A}{m_B + m_A}\vec{v} = \vec{0}$$
$$m_A\vec{v}'_A + m_B\vec{v}'_B = \vec{0}.$$

Hence vectors $\vec{v}'_A$ et $\vec{v}'_B$ are colinear.

Let

$$v'_A = \left|\vec{v}'_A\right| \quad \text{and} \quad v'_B = \left|\vec{v}'_B\right|$$

B. Mercier, *Simple Models for Understanding Nuclear Reactor Physics*,
https://doi.org/10.1007/978-3-031-98532-4

$$v'_B = \frac{m_A}{m_B} v'_A$$

Conservation of kinetic energy in the center of mass frame:

$$\frac{1}{2} m_A \left(\frac{m_B}{m_B + m_A} v \right)^2 + \frac{1}{2} m_B \left(\frac{m_A}{m_B + m_A} v \right)^2 = \frac{1}{2} m_A (v'_A)^2 + \frac{1}{2} m_B (v'_B)^2$$

$$\frac{m_A m_B}{m_B + m_A} v^2 = m_A (v'_A)^2 + m_B \left(\frac{m_A}{m_B} v'_A \right)^2 = m_A \frac{m_B + m_A}{m_B} (v'_A)^2$$

$$(v'_A)^2 = \left(\frac{m_B}{m_B + m_A} \right)^2 v^2$$

$$v'_A = \frac{m_B}{m_B + m_A} v \quad \text{and} \quad v'_B = \frac{m_A}{m_B} v'_A = \frac{m_A}{m_B + m_A} v$$

Remark A.1 Impulsions in the center of mass frame:

$$\vec{p}_A = \frac{m_A m_B}{m_B + m_A} \vec{v} = -\vec{p}_B$$
$$\vec{p}'_A = -\vec{p}'_B.$$

Then, for an elastic shock, in the center of mass frame, only the direction of both particles momentum changes. ■

Center of mass frame.

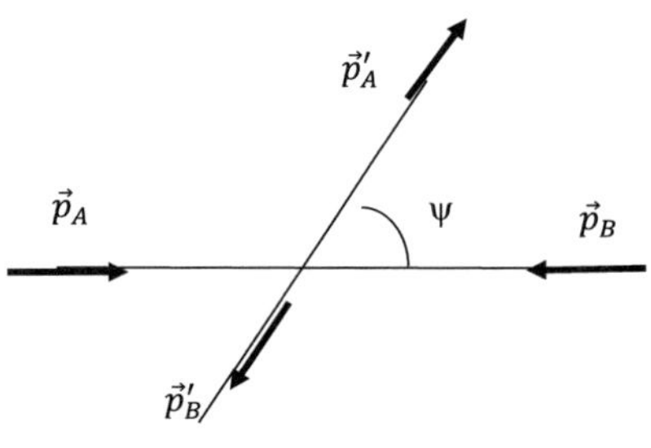

The conservation relations do not give the direction taken by the velocity $\vec{v}'_A$ of particle A after impact with its initial direction (nor by the velocity $\vec{v}'_B$ of the nucleus after impact, as it is collinear with it) in the center-of-mass reference frame.

In neutronics (A = neutron, B = nucleus), we'll assume that this direction is isotropic in the center of mass, i.e., the extremity of its unit vector $\vec{n}$ is uniformly distributed over the sphere of unit radius. Let $\psi \in [0, \pi]$ be the polar angle, i.e., the angle $\vec{v}, \vec{n}$ and $\varphi \in [0, 2\pi]$ the azimuthal angle around the axis carried by $\vec{v}$ we will take for true that $\mu = \cos$ is uniformly distributed in $[-1, +1]$ and φ in $[0, 2\pi]$.

We have

$$\vec{v}_A' = v_A' \begin{pmatrix} \cos\psi \\ \sin\psi \end{pmatrix} = \frac{m_B v}{m_B + m_A} \begin{pmatrix} \cos\psi \\ \sin\psi \end{pmatrix}$$

Hence, coming back to the laboratory frame, we have:

$$\vec{V}_A' = \vec{v}_G + \vec{v}_A' = \frac{m_A}{m_B + m_A}\vec{v} + \vec{v}_A'$$

$$\vec{V}_A' = V_A' \begin{pmatrix} \cos\theta \\ \sin\theta \end{pmatrix} = \frac{m_A v}{m_B + m_A} \begin{pmatrix} 1 \\ 0 \end{pmatrix} + \frac{m_B v}{m_B + m_A} \begin{pmatrix} \cos\psi \\ \sin\psi \end{pmatrix}$$

$V_A' \cos\theta = \frac{v}{m_B + m_A}(m_A + m_B \cos\psi)$ (note that if $m_A = m_B$, $\cos\theta \geq 0$ which shows anisotropy in the laboratory frame)

$$V_A' \sin\theta = \frac{v}{m_B + m_A} m_B \sin\psi$$

$$\left(V_A'\right)^2 = \left(\frac{v}{m_B + m_A}\right)^2 \left[m_A^2 + 2m_A m_B \cos\psi + m_B^2\right]$$

Replacing m_B by $x.m_A$ and introducing pre-impact kinetic energy $\left(E_A = \frac{1}{2} m_A v^2\right)$ and post-impact $\left(E_A' = \frac{1}{2} m_A \left(V_A'\right)^2\right)$ we get

$$E_A' = E_A \frac{1 + 2x\cos\psi + x^2}{(1+x)^2} = \frac{1}{2} E_A (1 + \gamma + (1-\gamma)\cos\psi) \tag{6.1}$$

where we let $\gamma = \frac{(x-1)^2}{(x+1)^2}$.

Indeed, $1 + \gamma + (1-\gamma)\cos\psi = \frac{(x+1)^2 + (x-1)^2}{(x+1)^2} + \left(\frac{(x+1)^2 - (x-1)^2}{(x+1)^2}\right)\cos\psi$.

Annex 2: Method for Obtaining the Diffusion Eq. (8.10)

We shall replace $\Phi(x, \Omega)$ by $\Phi_0(x) + 3\overrightarrow{\Omega} \cdot \vec{J}(x)$ in (8.5). However, since $\Phi_0 + 3\overrightarrow{\Omega} \cdot \vec{J}$ is just an approximation of Φ, we will obtain a residual

$$F = \Omega_i \frac{\partial}{\partial x_i}\left(\Phi_0 + 3\vec{\Omega} \cdot \vec{J}\right) + \Sigma\left(\Phi_0 + 3\vec{\Omega} \cdot \vec{J}\right) - S_0 - \frac{1}{4\pi}\Sigma_s \int_{S_2} \left(\Phi_0 + 3\overrightarrow{\Omega'} \cdot \vec{J}\right) \mathrm{d}\overrightarrow{\Omega'}$$

that will not be rigorously equal to zero.

Since we have 4 unknown functions $\Phi_0(x)$, $J_1(x)$, $J_2(x)$ et $J_3(x)$, we shall require that, for all $x \in V$

$$\int_{S_2} F(x, \Omega)\mathrm{d}\Omega = 0 \tag{A.1}$$

$$\int_{S_2} F(x, \Omega)\Omega_l \mathrm{d}\Omega = 0 \quad \text{for } l = 1, 2, 3 \tag{A.2}$$

To go a step further, we shall use the following relations (proved in Annex 3).

$$\int_{S_2} \Omega_i \mathrm{d}\Omega = 0 \quad \text{for } i = 1, 2, 3 \tag{A.3}$$

$$\int_{S_2} \Omega_i \Omega_k \mathrm{d}\Omega = \frac{4\pi}{3}\delta_{ik} \quad \text{for } i, k = 1, 2, 3 \tag{A.4}$$

where $\delta_{ik} = 1$ if $i = k$ and $= 0$ otherwise.

B. Mercier, *Simple Models for Understanding Nuclear Reactor Physics*, https://doi.org/10.1007/978-3-031-98532-4

We also have

$$\int_{S_2} \Omega_i \Omega_k \Omega_l \mathrm{d}\Omega = 0 \quad \text{for } i, k, l = 1, 2, 3 \tag{A.5}$$

From (A.3) we get

$$F = \Omega_i \frac{\partial}{\partial x_i}\left(\Phi_0 + 3\vec{\Omega}\cdot\vec{J}\right) + \Sigma\left(\Phi_0 + 3\vec{\Omega}\cdot\vec{J}\right) - S_0 - \Sigma_s \Phi_0$$

Reporting this relation in (A.1), we get

$$\int_{S_2} \left(\Omega_i \frac{\partial}{\partial x_i}\left(\Phi_0 + 3\vec{\Omega}\cdot\vec{J}\right) + \Sigma\left(\Phi_0 + 3\vec{\Omega}\cdot\vec{J}\right) - S_0 - \Sigma_s \Phi_0\right)\mathrm{d}\Omega = 0$$

and, using (A.3):

$$\int_{S_2} \left(3\Omega_i \Omega_k \frac{\partial}{\partial x_i} J_k + \Sigma\Phi_0 - S_0 - \Sigma_s \Phi_0\right)\mathrm{d}\Omega = 0$$

By using (A.4) we deduce that:

$$4\pi\left(\frac{\partial}{\partial x_k} J_k + \Sigma_0 \Phi_0 - S_0\right) = 0$$

or equivalently

$$\mathrm{div}\vec{J} + \Sigma_0 \Phi_0 = S_0.$$

Now injecting the expression obtained for F into (A.2), we obtain

$$\int_{S_2} \left(\Omega_i \frac{\partial}{\partial x_i}\left(\Phi_0 + 3\vec{\Omega}\cdot\vec{J}\right) + \Sigma\left(\Phi_0 + 3\vec{\Omega}\cdot\vec{J}\right) - S_0 - \Sigma_s \Phi_0\right)\Omega_l \mathrm{d}\Omega = 0$$

Then using relations (A.3) and (A.5), it comes:

$$\int_{S_2} \left(\Omega_i \frac{\partial}{\partial x_i} \Phi_0 + 3\Sigma\Omega_k J_k\right)\Omega_l \mathrm{d}\Omega = 0$$

i.e. using (A.4)

$$\frac{4\pi}{3}\left(\frac{\partial}{\partial x_l}\Phi_0 + 3\Sigma J_l\right) = 0$$

Thus, we obtain Fick's law (8.13) with $D = \frac{1}{3.\Sigma}$. ■

Annex 3: Proof of Formulae (A.3) to (A.5)

Let us explain the method we use to prove (A.3) and (A.4).

As an example, let us compute:

$$\int_{S_2} \Omega_1 \Omega_2 \mathrm{d}\Omega$$

Taking $\vec{\Omega} = \{\cos\varphi \cdot \sin\theta, \sin\varphi \cdot \sin\theta, \cos\theta\}$, we have

$$\int_{S_2} \Omega_1 \Omega_2 \mathrm{d}\Omega = \int_0^{2\pi} \mathrm{d}\varphi \int_0^{\pi} \sin\varphi \cos\varphi \cdot \sin^2\theta \sin\theta \mathrm{d}\theta = 0$$

indeed

$$\int_0^{2\pi} \sin\varphi \cos\varphi \mathrm{d}\varphi = \frac{1}{2}[-\cos 2\varphi]_0^{2\pi} = 0.$$

On the other hand,

$$\int_{S_2} (\Omega_1)^2 \mathrm{d}\Omega = \int_0^{2\pi} \mathrm{d}\varphi \int_0^{\pi} \cos^2\varphi \cdot \sin^2\theta \sin\theta \mathrm{d}\theta$$

$$= \frac{1}{2} \int_0^{2\pi} (1 + \cos 2\varphi) \mathrm{d}\varphi \int_{-1}^{+1} (1 - \mu^2) \mathrm{d}\mu = \frac{4\pi}{3}$$

B. Mercier, *Simple Models for Understanding Nuclear Reactor Physics*,
https://doi.org/10.1007/978-3-031-98532-4

To prove (A.5).

$$\int_{S_2} \Omega_i \Omega_k \Omega_l \mathrm{d}\Omega = 0$$

Let us take, as an example $i = k = l = 1$.
Since $\vec{\Omega} = \{\cos\varphi \cdot \sin\theta, \sin\varphi \cdot \sin\theta, \cos\theta\}$, we have

$$\Omega_i \Omega_k \Omega_l = \cos^3\varphi \cdot \sin^3\theta,$$

and then

$$\int_{S_2} \Omega_i \Omega_k \Omega_l \mathrm{d}\Omega = \int_0^{2\pi} \mathrm{d}\varphi \int_0^{\pi} \cos^3\varphi \cdot \sin^3\theta \sin\theta \mathrm{d}\theta = 0$$

Indeed

$$\int_0^{2\pi} \cos^3\varphi \mathrm{d}\varphi = \int_{-\frac{\pi}{2}}^{\frac{\pi}{2}} \cos^3\varphi \mathrm{d}\varphi + \int_{\frac{\pi}{2}}^{\frac{3\pi}{2}} \cos^3\varphi \mathrm{d}\varphi = \int_{-\frac{\pi}{2}}^{\frac{\pi}{2}} \left[\cos^3\varphi + \cos^3(\pi + \varphi)\right] \mathrm{d}\varphi = 0$$

The other cases can be computed in a similar way.

Annex 4: Method to Obtain the Fourier Boundary Condition (8.11)

We saw for the Boltzmann equation that the neutron current flowing through $x \in \Gamma$ the small surface element $\mathrm{d}S$ orthogonal to the vector $\vec{n}(x)$ was $\vec{\Omega} \cdot \vec{n}\Phi\mathrm{d}S$.

Integrating for all $\Omega \in \boldsymbol{S_2}$ we obtain

$$\mathrm{d}S \int\limits_{S_2} \vec{\Omega} \cdot \vec{n}\Phi\mathrm{d}\Omega = \mathrm{d}S \int\limits_{S_2} \vec{\Omega} \cdot \vec{n}(\Phi_0 + 3\Omega_k J_k)\mathrm{d}\Omega = 4\pi\vec{J} \cdot \vec{n}\mathrm{d}S$$

which shows that $\vec{J}$ can be interpreted as the neutron current.

In a similar way to what we did for the Boltzmann equation in 1D (see Exercise 8.1), it is logical to separately evaluate J_+ and J_- defined by

$$J_+ = \frac{1}{4\pi} \int\limits_{\vec{\Omega}\cdot\vec{n}>0} \vec{\Omega} \cdot \vec{n}\Phi\mathrm{d}\Omega$$

$$J_- = -\frac{1}{4\pi} \int\limits_{\vec{\Omega}\cdot\vec{n}<0} \vec{\Omega} \cdot \vec{n}\Phi\mathrm{d}\Omega.$$

Introducing axes connected to the normal vector $\overrightarrow{n}$, θ will denote the polar angle et φ the azimuth angle, we will still have in this local frame

$$\vec{\Omega} = \{\cos\varphi \cdot \sin\theta, \sin\varphi \cdot \sin\theta, \cos\theta\}$$

The half space $\vec{\Omega} \cdot \vec{n} > 0$ corresponds to $0 \leq \theta < \frac{\pi}{2}$ and we have

$$J_+ = \frac{1}{4\pi} \int\limits_0^{2\pi} \mathrm{d}\varphi \int\limits_0^{\frac{\pi}{2}} \cos\theta(\Phi_0 + 3(\cos\varphi \cdot \sin\theta J_1 + \sin\varphi \cdot \sin\theta J_2 + \cos\theta J_3)) \sin\theta\mathrm{d}\theta$$

B. Mercier, *Simple Models for Understanding Nuclear Reactor Physics*,
https://doi.org/10.1007/978-3-031-98532-4

Note that $J_k, k = 1, 2, 3$ are now related to the local reference frame so that $J_3 = \vec{J}.\vec{n}$. Integrating w.r.t. φ we see that the coefficients of J_1 and J_2 vanish. Here is what remains:

$$J_+ = \frac{1}{2}\int_0^{\frac{\pi}{2}} \cos\theta(\Phi_0 + 3\cos\theta J_3)\sin\theta \mathrm{d}\theta = \frac{1}{2}\int_0^1 \mu(\Phi_0 + 3\mu J_3)\mathrm{d}\mu = \frac{1}{4}\Phi_0 + \frac{1}{2}J_3$$

i.e.

$$J_+ = \frac{1}{4}\Phi_0 - \frac{D}{2}\frac{\partial}{\partial n}\Phi_0$$

In the same way, we would get:

$$J_- = \frac{1}{4}\Phi_0 - \frac{1}{2}J_3 = \frac{1}{4}\Phi_0 + \frac{D}{2}\frac{\partial}{\partial n}\Phi_0$$

Then, if no neutron enters V, it is logical to take (8.11), the interpretation of which is $J_- = 0$, as boundary condition.

Appendix
Solution to Exercises

Solution to Exercise 2.1

200 MeV corresponds to 3.21 10^{-11} J.

To produce 3.8 10^9 J/s we need $\frac{3.8\ 10^9}{3.21\ 10^{-11}} = 1.19\ 10^{20}$ fissions per second.

In each fuel cm^3 we should have, in average, $\frac{1.19\ 10^{20}}{1.11\ 10^7} = 1.07\ 10^{13}$ fissions per second, that is 8.87 10^{12} fissions due to *thermal neutrons* per cm^3 per second.

Let Φ_{th} denote the thermal neutron flux in the fuel, from what we have seen above, we deduce that $\Sigma_f\,\Phi_{th}$ is the number of fissions due to *thermal neutrons* per cm^3 per second. Hence, we get:

$$\Phi_{th} = \frac{8.87\ 10^{12}}{\Sigma_f} = \frac{8.87\ 10^{12}}{0.24} = 3.69\ 10^{13}\ \text{n/cm}^2\text{/s} \quad \blacksquare$$

Solution to Exercise 3.1

Let $N_A = 0.6023\ 10^{24}$ be the Avogadro number.

Since the mass of a UO_2 (resp. PuO_2) molecule is equal to 270 g (resp. 271g), the number of molecules in 1 cm^3 of MOX is approximately 10.3/270. $N_A = 0.02298$ molecules/(b.cm).

With obvious notations we get $\Sigma_{f\text{MOX}} = N_{\text{U}}.\sigma_f^8 + N_{\text{Pu}}.\sigma_f^9$, $\Sigma_{c\text{MOX}} = N_{\text{U}}.\sigma_c^8 + N_{\text{Pu}}.\sigma_c^9$, and

	$e = 11\%$	$e = 10\%$
N_{U} (at/(b.cm)	0.02045	0.02068
N_{Pu} at/(b.cm)	0.002527	0.002298
$\Sigma_{f\text{MOX}}$ (cm^{-1})	0.005620	0.005193
$\Sigma_{c\text{MOX}}$ (cm^{-1})	0.007651	0.007582

Note that 1 at/(b.cm) = 10^{24} at/cm^3

B. Mercier, *Simple Models for Understanding Nuclear Reactor Physics*, https://doi.org/10.1007/978-3-031-98532-4

For Sodium, we have in 1 cm^3

$$N_{\text{Na}} = 0.97/23 \,.\, N_A = 0.0254\,\text{at}/(\text{b.cm}) \text{ then } \Sigma_{c\text{Na}} = 0.000254\,\text{cm}^{-1}.$$

For the cladding, we have in 1 cm^3

$$N_{\text{Fe}} = 7.8/56.\, N_A = 0.0834\,\text{at}/(\text{b.cm}) \text{ then } \Sigma_{c\text{Fe}} = 0.00834\,\text{cm}^{-1}.$$

To apply formula (2.1) we need to compute Σ_f and $\Sigma_a = \Sigma_f + \Sigma_c$ at the core level.

We have

$$\Sigma_f = \omega_1.\Sigma_{f\text{MOX}} = \begin{Bmatrix} 0.003662\,\text{cm}^{-1} & \text{when } e = 11\% \\ 0.003383\,\text{cm}^{-1} & \text{when } e = 10\% \end{Bmatrix}$$

$$\Sigma_a = \omega_1.\Sigma_{a\text{MOX}} + \omega_2.\Sigma_{a\text{clad}} + \omega_3.\Sigma_{c\text{Na}} = \begin{Bmatrix} 0.01038\,\text{cm}^{-1} \text{ when } e = 11\% \\ 0.01006\,\text{cm}^{-1} \text{ when e } = 10\% \end{Bmatrix}$$

$$k_\infty = \begin{Bmatrix} 1.0262 \text{ when e } = 11\% \\ 0.9787 \text{ when e } = 10\% \end{Bmatrix}$$

We conclude that the core cannot diverge if the enrichment is equal to 10% or smaller, even if the core is very large. Actually on Superphenix (see [18, p. 191]) the ^{239}Pu equivalent enrichment of the MOX was more like 15%, which gives $k_\infty = 1.19$, but the core could not be very large and, as for all fast neutron reactors, the leakage was significant. ■

Solution to Exercise 5.1

With pure water $f = \frac{\Sigma_{\text{fuel}}/3}{\Sigma_{\text{fuel}}/3+2\Sigma_{H_2O}/3} = \frac{0.6}{0.6+2*0.016} = 0.949$.

With 1000 ppm borated water, Σ_{H_2O} is replaced by $\Sigma_* = 0.046\,\text{cm}^{-1}$ hence $f = 0.867$ which induces a 0.082 reactivity gap i.e. 8200 pcm. ■

Solution to Exercise 5.2

1. We find $p = \frac{100}{133}\, f = \frac{78}{97}\, \eta = 2.4\frac{60}{78}\, \varepsilon = \frac{150}{144}\, \mathcal{L}_f = \frac{133}{150}\, \mathcal{L}_t = \frac{97}{100}$.

   The product of those 6 factors is then

$$\frac{150}{144} \times \frac{133}{150} \times \frac{100}{133} \times \frac{97}{100} \times \frac{78}{97} \times 2.4 \times \frac{60}{78} = 2.4 \times \frac{60}{144} = 1$$

2. We have $k_\infty = \varepsilon\,.\,p\,.\,f\,.\,\eta = \frac{150}{144} \times \frac{100}{133} \times \frac{78}{97} \times 2.4 \times \frac{60}{78} = 1.162$.
3. The fission probability is $P_f = \frac{60}{78}$.

   Let e denote the fuel enrichment,

$$P_f = \frac{e\sigma_f^5}{e\sigma_a^5 + (1-e)\sigma_a^8}$$

i.e.

$$\left(e\sigma_a^5 + (1-e)\sigma_a^8\right)P_f = e\sigma_f^5$$
$$\sigma_a^8 P_f = e\left[\sigma_f^5 - P_f\left(\sigma_a^5 - \sigma_a^8\right)\right]$$

Hence $e = \frac{\sigma_a^8 P_f}{\sigma_f^5 - P_f\left(\sigma_a^5 - \sigma_a^8\right)} = 3.5\%$ ■

Solution to Exercise 7.1

We have $E = E_0 e^{-u}$ then $E = 9.11\,\text{keV}$ for $u = 7$; then $E = 3.06\,\text{eV}$ when $u = 15$.

Therefore the lethargy interval (7, 15) contains the ^{238}U traps. ■

Solution to Exercise 7.2

We have

$$\mathbb{E}\text{w} = \int_0^{\varepsilon} w \frac{e^{-w}}{1-\gamma} dw$$

Let $u = w$ and $v = -\frac{e^{-w}}{1-\gamma}$ we have (note that $e^{-\varepsilon} = \gamma$)

$$\begin{aligned}
\mathbb{E}w &= \int_0^{\varepsilon} uv' dw = [uv]_0^{\varepsilon} - \int_0^{\varepsilon} u'v dw = \left[-w\frac{e^{-w}}{1-\gamma}\right]_0^{\varepsilon} + \int_0^{\varepsilon} \frac{e^{-w}}{1-\gamma} dw \\
&= -\varepsilon\frac{e^{-\varepsilon}}{1-\gamma} + \frac{1}{1-\gamma}\int_0^{\varepsilon} e^{-w} dw = -\varepsilon\frac{\gamma}{1-\gamma} + \frac{1}{1-\gamma}\left[-e^{-w}\right]_0^{\varepsilon} \\
&= -\frac{\gamma\varepsilon}{1-\gamma} + 1
\end{aligned}$$

■

Solution to Exercise 7.3

1. Let $e = 0.037$ be the enrichment.

   The molar mass of a UO_2 molecule is $(235\,e + 238\,(1-e) + 32) \cong 270$.
   In 1 cm^3 of fuel there is 11/270 UO_2 moles.
   Let $N_A = 0.6023\ 10^{24}$ be the Avogadro number. In each fuel cm^3, we have

$$11/270\,.\,N_A = 0.0245\ 10^{24}\ UO_2 \text{ molecules.}$$

Since 1 barn is 10^{-24} cm^2, we deduce $N_U = 0.0245$ at/(barn.cm), $N_8 = 0.02356$ at/(barn.cm), and

$$N_5 = 0.000907 \,\text{at}/(\text{barn.cm}).$$

2. We neglect the oxygen absorption. We have, with obvious notations, $\Sigma_{aU5} = N_5\,\sigma_5$ $\Sigma_{aU8} = N_8\,\sigma_8$ and $\Sigma_{a\text{fuel}} = \Sigma_{aU5} + \Sigma_{aU8} = 0.681\ \text{cm}^{-1}$.
3. Proceeding in a similar way, we find $N_H = 0.06692/(\text{barn.cm})$, $\Sigma_{am} = 0.0215\ \text{cm}^{-1}$.
4. We have $f = \frac{\omega \Sigma_{a\text{fuel}}}{\omega \Sigma_{a\text{fuel}} + (1-\omega)\Sigma_{am}} = 0.94$. ■

Solution to Exercise 7.4

Proceeding as in the solution of the previous exercise, we find

$$N_G = N_A . \frac{1.6}{12} = 0.0803\ 10^{24}\ \text{at/cm}^3; \quad \Sigma_{cG} = N_G . \sigma_{cG} = 0.000273\,\text{cm}^{-1}.$$

■

Solution to Exercise 8.1

1. We note that

$$\int_{-1}^{1} 3\mu' J \,\mathrm{d}\mu' = J(z) \int_{-1}^{1} 3\mu' \mathrm{d}\mu' = 0$$

so that

$$F = \mu \frac{\partial}{\partial z}(\Phi_0 + 3\mu J) + \Sigma_0 \Phi_0 + 3\mu \Sigma J - S_0$$

2. We have $\int_{-1}^{1} \left(\mu \frac{\partial}{\partial z}(\Phi_0 + 3\mu J) + \Sigma_0 \Phi_0 + 3\mu \Sigma J - S_0\right) \mathrm{d}\mu = 0$ and then

$$\left[\mu^3\right]_{-1}^{+1} \frac{\mathrm{d}}{\mathrm{d}z} J(z) + 2\Sigma_0 \Phi_0 = 2S_0$$

Which proves the desired result.
3. We have

$$\int_{-1}^{1} \left[\mu \frac{\partial}{\partial z}(\Phi_0 + 3\mu J) + \Sigma_0 \Phi_0 + 3\mu \Sigma J - S_0\right] \mu \mathrm{d}\mu = 0 \quad \text{and then}$$

$$\int_{-1}^{1} \left[\mu^2 \frac{\partial \Phi_0}{\partial z} + \Sigma\ 3\mu^2 J\right] \mathrm{d}\mu = 0$$

Which proves the desired result.

4. We have

$$J_+(z) = \frac{1}{2}\int_0^1 (\Phi_0(z) + 3\mu J(z))\mu \mathrm{d}\mu$$
$$= \frac{1}{2}\left[\frac{\mu^2}{2}\right]_0^1 \Phi_0(z) + \frac{1}{2}\left[\mu^3\right]_0^1 J(z) = \frac{1}{4}\Phi_0(z) + \frac{1}{2}J(z)$$

and similarly

$$J_-(z) = -\frac{1}{2}\int_{-1}^0 (\Phi_0(z) + 3\mu J(z))\mu \mathrm{d}\mu$$
$$= -\frac{1}{2}\left[\frac{\mu^2}{2}\right]_{-1}^0 \Phi_0(z) - \frac{1}{2}\left[\mu^3\right]_{-1}^0 J(z) = \frac{1}{4}\Phi_0(z) - \frac{1}{2}J(z)$$

Which proves the desired result.

5. The boundary condition (8.15) implies that

$$J_+(0) = \frac{1}{4}A$$

$$J_-(L) = 0$$

hence the desired result. ■

Solution to Exercise 8.2

1. We have

$$\Phi''(z) = \beta^2 \Phi(z)$$

Now it's obvious that $e^{\beta z}$ and $e^{-\beta z}$ are two independent solutions of this second-order linear differential equation with constant coefficients. By half-sum and half-difference, we see that $ch(\beta z)$ and $sh(\beta z)$ are also two independent solutions. The general solution is therefore a linear combination of these two solutions, hence the result.

2. We have

$$\Phi'(z) = \beta a\, sh(\beta z) + \beta b\, ch(\beta z)$$

And then (see Annex 4):

$$J_+(z) = \frac{1}{4}\left(\Phi(z) - 2D\Phi'(z)\right) = \frac{1}{4}((a - 2D\beta b)\, ch(\beta z) + (b - 2D\beta a)\, sh(\beta z))$$
$$J_-(z) = \frac{1}{4}\left(\Phi(z) + 2D\Phi'(z)\right) = \frac{1}{4}((a + 2D\beta b)\, ch(\beta z) + (b + 2D\beta a)\, sh(\beta z))$$

Using that, in $z = L$, we have $J_-(L) = 0$, we find

$$(a + 2D\beta b)\, ch(\beta L) + (b + 2D\beta a)\, sh(\beta L) = 0$$

that is a first relation between a et b.

To obtain a second one, we use that, in $z = 0$, $J_+(0) = A$ is given. Thus, we have

$$a/4 - D\beta b/2 = A.$$

A given, we can find a and b by solving the 2×2 system.

$$\begin{pmatrix} ch(\beta L) + 2D\beta\, sh(\beta L) & sh(\beta L) + 2D\beta ch(\beta L) \\ 1/4 & -D\beta/2 \end{pmatrix} \begin{pmatrix} a \\ b \end{pmatrix} = \begin{pmatrix} 0 \\ A \end{pmatrix}$$

From a and b we can compute $\Phi(z)$, $J_+(z)$ et $J_-(z)$.

3. From the first equation of our 2×2 system, we obtain:

$$a = -\frac{sh(\beta L) + 2D\beta ch(\beta L)}{ch(\beta L) + 2D\beta sh(\beta L)} b = -\frac{th(\beta L) + 2D\beta}{1 + 2D\beta th(\beta L)} b$$

Reporting in the second equation we find

$$\left(\frac{th(\beta L) + 2D\beta}{1 + 2D\beta th(\beta L)} + 2D\beta\right) b = -4A$$

this is a nonlinear equation that we solve by the trial and errors method.

Numerical application.

We find $\beta = 0.272\,\text{cm}^{-1}$, $b = -174.36\,\text{n/cm}^2/\text{s}$; $a = +174.36\,\text{n/cm}^2/\text{s} = \Phi(0)$

$$\Phi'(0) = \beta b = -47.465$$
$$J_-(0) = 37.184\,\text{n/cm}^2/\text{s}$$
$$\Phi'(L) \cong 0 \quad \text{and then } J(L) \cong 0 \text{ so that } J_+(L) \cong 0$$

So $50\,\text{n/cm}^2/\text{s}$ are incoming in our slab of lightwater, but only $37.184\,\text{n/cm}^2/\text{s}$ are outgoing on the side $z = 0$. As we shall see, the missing neutrons are absorbed by light water.

Note that $b = -a$ so that $\Phi(x) = a\mathrm{e}^{-\beta x}$, then Φ is decreasing exponentially.

4. Integrating (8.20) from 0 to L we find

$$J(L) - J(0) + \int_0^L \Sigma_a \cdot \Phi \mathrm{d}z = 0$$

As $J(L) \cong 0$ we find

$$\text{Absorption/cm}^2/\text{s} = \int_0^L \Sigma_a.\Phi \mathrm{d}z = J(0) = J_+(0) - J_-(0) = 12.8\,\text{abs/cm}^2/\text{s}$$

A layer of light water is then a good neutron shielding, and an acceptable reflector. ■

Solution to Exercise 8.3

1. Assuming $\Sigma_a = 0$, from (8.20) we get

$$\frac{\mathrm{d}J}{\mathrm{d}z} = 0$$

hence $J(0) = J(L) = J$; on the other hand we have

$$J = J(z) = -D\Phi'(z)$$

hence

$$\Phi(z) = b - \frac{J}{D}z.$$

On the other hand, we have, as above:

$$J_+(z) = \frac{1}{4}\big(\Phi(z) - 2D\Phi'(z)\big)$$
$$J_-(z) = \frac{1}{4}\big(\Phi(z) + 2D\Phi'(z)\big)$$

Using that, in $z = L$, $J_-(L) = 0$, we find

$$b = \frac{J}{D}L + 2J$$

We deduce that, in $z = 0$

$$J_-(0) = \frac{1}{4}\frac{J}{D}L$$
$$J_+(0) = J + \frac{1}{4}\frac{J}{D}L = J\left(1 + \frac{L}{4D}\right)$$

and of course, in $z = L$, we have $J_+(L) = J$.

2. We have $\sigma \cong 10\,b$ and $N_{D20} = 0.0332\ 10^{24}$ molecules/cm^3.

We deduce $\Sigma \cong 0.332\,\text{cm}^{-1}$ and then $D \cong \frac{1}{3\Sigma} \cong 1\,\text{cm}$.

Since $\frac{L}{D} \cong 50$, we get $J \cong \frac{50}{13.5} = 3.7$ and $J_-(0) = A - J = 50 - 3.7 = 46.3\,\text{n/cm}^2/\text{s}$.

So, heavy water is a better reflector than light water. On the other hand, it is not good for neutron shielding since $J_+(L) = J + J_-(L) = J = 3.7\,\text{n/cm}^2/\text{s}$. So the outgoing current is significant. ■

Solution to Exercise 9.1

Let $x = x_0 \frac{r}{R}$ so that

$$\frac{\mathrm{d}}{\mathrm{d}r}\Phi_0(r) = \frac{\mathrm{d}x}{\mathrm{d}r}\frac{\mathrm{d}}{\mathrm{d}x}J_0 = \frac{x_0}{R}J_0'(x)$$
$$\frac{\mathrm{d}^2}{\mathrm{d}r^2}\Phi_0(r) = \left(\frac{x_0}{R}\right)^2 J_0''$$

From the formula giving Laplace's operator in polar coordinates, we have

$$\Delta\Phi_0 = \frac{\mathrm{d}^2}{\mathrm{d}r^2}\Phi_0(r) + \frac{1}{r}\frac{\mathrm{d}}{\mathrm{d}r}\Phi_0(r) = \left(\frac{x_0}{R}\right)^2 J_0''(x) + \frac{x_0}{R}\frac{1}{x}\frac{x_0}{R}J_0'(x)$$
$$= -\left(\frac{x_0}{R}\right)^2 J_0(x) = -\mu_0\Phi_0$$

On the other hand, from the definition of x_0 we have

$$\Phi_0(R) = J_0(x_0) = 0$$

Φ_0 satisfies also the boundary condition on the disk edge. Thus we have proved that μ_0 is an eigenvalue of our problem.

Since the associated eigenmode Φ_0 is positive on the disk, it results from the Krein–Rutman theorem (see [8]) that it is the smallest eigenvalue. Then Φ_0 is the fundamental eigenmode.

2. Let $u_0 = u_0(x, y, z)$ be the eigenmode associated to λ_0.

For fixed x and y we can develop function $z \to u_0(x, y, z)$ in Fourier series, i.e.:

$$u_0(x, y, z) = \sum_{n\geq 0} v_n(x, y)\sin\left(n\pi\frac{z}{H}\right)$$

From the definition of u_0 and λ_0 we have

$$0 = \Delta u_0 + \lambda_0 u_0 = \sum_{n \geq 0} \left(\frac{\partial^2}{\partial x^2} v_n + \frac{\partial^2}{\partial y^2} v_n + \left(\lambda_0 - \left(\frac{n\pi}{H} \right)^2 \right) v_n \right) \sin\left(n\pi \frac{z}{H} \right)$$

A Fourier series vanishes if and only if all its coefficients vanish.
Then either $v_n = 0$ or

$$\frac{\partial^2}{\partial x^2} v_n + \frac{\partial^2}{\partial y^2} v_n + \left(\lambda_0 - \left(\frac{n\pi}{H} \right)^2 \right) v_n = 0$$

which means that v_n is an eigenfunction of minus the Laplace operator vanishing on the edge of the disk of R so that

$$\lambda_0 = \left(\frac{n\pi}{H} \right)^2 + \mu_n$$

where μ_n is the eigenvalue associated to v_n, so that μ_n is positive.
The smallest choice corresponds to $n = 1$ so that

$$\lambda_0 = \left(\frac{\pi}{H} \right)^2 + \mu_0$$

Hence, the result.

3. Our constraint implies that $R = R(H)$ so we let

$$f(H) = \left(\frac{x_0}{R(H)} \right)^2 + \left(\frac{\pi}{H} \right)^2 = x_0^2 \frac{\pi H}{V} + \frac{\pi^2}{H^2}$$

The optimality condition will be

$$0 = f'(H) = x_0^2 \frac{\pi}{V} - 2\frac{\pi^2}{H^3}$$

When $H = \frac{\pi\sqrt{2}}{x_0} R$, we have $H^2 = 2\frac{\pi^2}{x_0^2} R^2$

$$H^3 = HH^2 = 2\frac{\pi}{x_0^2} V$$

So that we get

$$f'(H) = x_0^2 \frac{\pi}{V} - 2\pi^2 \frac{x_0^2}{2\pi V} = 0$$

Application

We find $\lambda_0 = 2.819\,\mathrm{m}^{-2}$ and $M^2 = 0.007239\,\mathrm{m}^2 = 72.39\,\mathrm{cm}^2$. ■

Solution to Exercise 9.2

Let $\vec{J} = -D\vec{\nabla}\Phi$ denote the neutron current, from (9.13) we have

$$\mathrm{div}\vec{J} + \Sigma_0\Phi = \frac{k_\infty}{k}\Sigma_0\Phi \quad \text{for } x \in V$$

Hence by integration by parts, we have leakage + absorption $= \frac{k_\infty}{k}$. absorption. Incidentally, we have proved that the non-leakage probability is $\frac{\text{absorption}}{\text{leakage+absorption}}$ ■

Solution to Exercise 9.3

1. For $a = H$ we have

$$-M_B^2\Phi'' + \Phi = \frac{k_B}{k}\Phi$$

So that $\Phi(z) = \sin\frac{\pi z}{H}$ and

$$\frac{k_B}{k} = 1 + M_B^2\left(\frac{\pi}{H}\right)^2$$

We compute successively $k_B = 1.00849$, $M_B^2 = 49\,\mathrm{cm}^2$, $\left(\frac{\pi}{H}\right)^2 = 6.1685\,\mathrm{cm}^{-2}$ then $k = 1.00545$.

For $a = 0$, that corresponds to fully inserted absorbing rods, in a similar way, we find

$$\frac{k_H}{k} = 1 + M_H^2\left(\frac{\pi}{H}\right)^2$$

$k_H = 0.98$, $M_H^2 = 47.616\,\mathrm{cm}^2$ then $k = 0.97712$.
The reactivity gap between both configurations is then equal to 2748 pcm.

2. From Fick's law $J(z) = -D\Phi'(z)$ so that Φ is differentiable, then continuous.

Now $J(z)$ is the neutron current. It must be continuous (otherwise, we would loose neutrons at the interface $z = a$) hence the result.

3. The general solution of (9.28) is a linear combination of $\sin(\alpha z)$ and $\cos(\alpha z)$. Now the boundary condition (9.26) eliminates the latter. So that the result holds for $0 < z < a$.

For $a < z < H$, the general solution of (9.29) is a linear combination of $sh(\beta z)$ and $ch(\beta z)$.

The addition formula

$$sh(\beta(H-z)) = sh(\beta H)ch(\beta z) - sh(\beta z)ch(\beta H)$$

Shows that this is the case for $sh(\beta(H-z))$ and it vanishes in $z = H$.

4. We first calculate:

$$\Phi'(z) = \begin{Bmatrix} \alpha A\cos(\alpha z) & \text{for } z < a \\ -\beta Bch(\beta(H-z)) & \text{for } z > a \end{Bmatrix}$$

By using continuity of Φ and J in $z = a$, we obtain

$$A\sin(\alpha a) = Bsh(\beta(H-a))$$
$$\alpha A\cos(\alpha a) = -\beta Bch(\beta(H-a))$$

Hence the result by dividing each side of the second equation by each side of the first equation.

5. We find

a	k
0	0.97712
100	0.98113
200	1.00034
300	1.00433
400	1.00545

6. With this new choice for β (which requires $k < k_H$), (9.29) is replaced by

$$-\Phi'' = \beta^2\Phi$$

hence the result.

7. We have

$$\Phi'(z) = \begin{Bmatrix} \alpha A\cos(\alpha z) & \text{for } z < a \\ -\beta B\cos(\beta(H-z)) & \text{for } z > a \end{Bmatrix}$$

Applying continuity of Φ and J in $z = a$, we obtain

$$A\sin(\alpha a) = B\sin(\beta(H-a))$$
$$\alpha A\cos(\alpha a) = -\beta B\cos(\beta(H-a))$$
$$\frac{\alpha}{\text{tg}(\alpha a)} + \frac{\beta}{\text{tg}(\beta(H-a))} = 0$$

8. We find $k = 0.97775$.

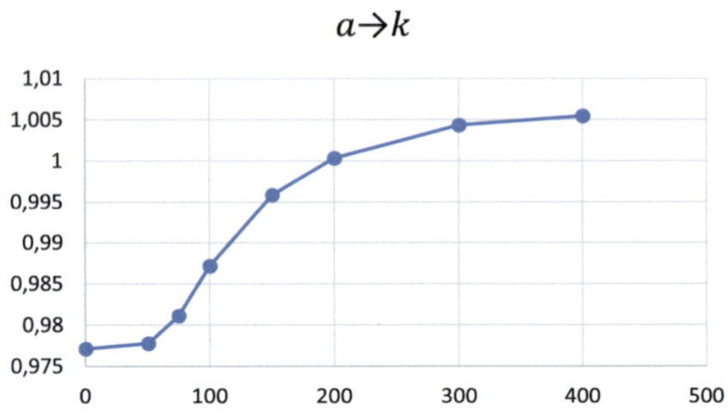

9. We find the above curve.

We observe the reactivity drops quite slowly when the absorbing bars are introduced, then faster when the core's mid-plane is reached, and finally more slowly.

10. We find the curve below.

We observe that both Φ and Φ' are continuous and that Φ'' is negative for $z < a$ and positive for $z > a$.

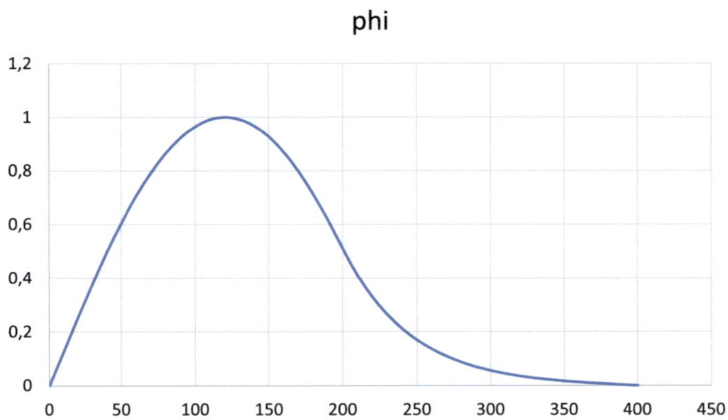

■

Solution to Exercise 9.4

1. If we set

$$\alpha = \sqrt{(k_\infty - 1)/M^2}$$

(9.30) proves that $\Phi(z) = A\cos(\alpha z) + B\sin(\alpha z)$.
(9.31) proves that $B = 0$.
We can choose $A = 1$.

2. We know from Exercise 8.1 that

$$J_-(z) = \Phi(z) + 2D\Phi'(z)$$
$$J_+(z) = \Phi(z) - 2D\Phi'(z)$$

So that

$$J_-(z) = \cos(\alpha z) - 2D\alpha\sin(\alpha z) \quad \text{and}$$
$$J_+(z) = \cos(\alpha z) + 2D\alpha\sin(\alpha z)$$

From (9.32) we have

$$\cos(\alpha a) - 2D\alpha\sin(\alpha a) = \theta(\cos(\alpha a) + 2D\alpha\sin(\alpha a))$$

So that

$$(1-\theta)\cos(\alpha a) = (1+\theta)2D\alpha\sin(\alpha a)$$

Hence, we get the desired result.

3. We find

$$\alpha = 0.00627\,\text{cm}^{-1} \quad \text{for}\,\theta = 0$$
$$\alpha = 0.00735\,\text{cm}^{-1} \quad \text{for}\,\theta = 0.7437$$
$$\alpha = 0.00777\,\text{cm}^{-1} \quad \text{for}\,\theta = 0.926$$

4. Since $k_\infty = 1 + \alpha^2 M^2$ we get

$$k_\infty = 1.03023 \quad \text{for}\,\theta = 0$$
$$k_\infty = 1.02705 \quad \text{for}\,\theta = 0.7437$$
$$k_\infty = 1.01969 \quad \text{for}\,\theta = 0.926$$

The benefit of a light water reflector is then + 318 pcm.
The benefit of a heavy water reflector is then + 1054 pcm. ■

Solution to Exercise 10.1

1. (the proof is quite long but interesting).

We can rewrite (10.8) and (10.9) as follows:

$$-\frac{\partial^2}{\partial z^2}\psi + a\psi + b\varphi = 0, \quad 0 < z < L$$
$$-\frac{\partial^2}{\partial z^2}\varphi + c\psi + \mathrm{d}\varphi = 0 \quad 0 < z < L$$

i.e.

$$-\frac{\partial^2}{\partial z^2}\begin{pmatrix}\psi\\ \varphi\end{pmatrix}+A\begin{pmatrix}\psi\\ \varphi\end{pmatrix}=0 \quad \text{with}$$

$$A=\begin{pmatrix}a & b\\ c & d\end{pmatrix}=\begin{pmatrix}M_1^{-2}(1-k_{\infty,1}) & -Qk_{\infty,2}M_1^{-2}\\ -Q^{-1}M_2^{-2} & (1+P)M_2^{-2}\end{pmatrix}$$

Proposition *If* $\frac{k_{\infty,2}}{1+P}+k_{\infty,1}>1$, *the eigenvalues* λ *and* μ *of* A *have opposite signs.*

Proof If x is an eigenvalue of A, there exists $\begin{pmatrix}\alpha\\ \beta\end{pmatrix}$ s.t.

$$\begin{pmatrix}M_1^{-2}(1-k_{\infty,1}) & -Qk_{\infty,2}M_1^{-2}\\ -Q^{-1}M_2^{-2} & (1+P)M_2^{-2}\end{pmatrix}\begin{pmatrix}\alpha\\ \beta\end{pmatrix}=x\begin{pmatrix}\alpha\\ \beta\end{pmatrix}$$

$$\begin{pmatrix}1-k_{\infty,1} & -Qk_{\infty,2}\\ -Q^{-1} & (1+P)\end{pmatrix}\begin{pmatrix}\alpha\\ \beta\end{pmatrix}=x\begin{pmatrix}M_1^2\alpha\\ M_2^2\beta\end{pmatrix}$$

$$\begin{pmatrix}1-k_{\infty,1}-xM_1^2 & -Qk_{\infty,2}\\ -Q^{-1} & 1+P-xM_2^2\end{pmatrix}\begin{pmatrix}\alpha\\ \beta\end{pmatrix}=\begin{pmatrix}0\\ 0\end{pmatrix}$$

Then we have

$$(1-k_{\infty,1}-xM_1^2)(1+P-xM_2^2)=k_{\infty,2} \tag{10.10}$$

i.e.

$$x^2M_1^2M_2^2-x\big(M_1^2(1+P+M_2^2(1-k_{\infty,1})\big)+(1-k_{\infty,1})(1+P)-k_{\infty,2}=0$$

Which proves the result.

We now assume that $\frac{k_{\infty,2}}{1+P}+k_{\infty,1}>1$ (means that the core would be critical if there were no leaks)

Let us denote by λ the negative eigenvalue and by μ the positive eigenvalue. Note that in such a case matrix A is diagonalizable.

Let $D=\begin{pmatrix}\lambda & 0\\ 0 & \mu\end{pmatrix}$ there exists a 2×2 matrix P s.t.

$$A=P^{-1}DP$$

and then

$$-\frac{\partial^2}{\partial z^2}P^{-1}P\begin{pmatrix}\psi\\ \varphi\end{pmatrix}+P^{-1}DP\begin{pmatrix}\psi\\ \varphi\end{pmatrix}=0$$

Let $\begin{pmatrix} u \\ v \end{pmatrix} = P\begin{pmatrix} \psi \\ \varphi \end{pmatrix}$ we have

$$-\frac{\partial^2}{\partial z^2}\begin{pmatrix} u \\ v \end{pmatrix} + D\begin{pmatrix} u \\ v \end{pmatrix} = 0$$

i.e.

$$\begin{aligned} -\frac{\partial^2}{\partial z^2}u + \lambda u = 0, \quad 0 < z < H \\ -\frac{\partial^2}{\partial z^2}v + \mu v = 0 \quad 0 < z < H \\ u(0) = u(H) = v(0) = v(H) = 0 \end{aligned}$$

Hence:

$$\begin{aligned} u &= \sin z\sqrt{-\lambda} \\ v &= \exp(\pm z\sqrt{\mu}) \end{aligned}$$

Now, the boundary conditions imply $v = 0$ and $\sqrt{-\lambda} = \pi/H$ so that $\lambda = -\lambda_0 = -\pi^2/H^2$.

Let $P^{-1} = \begin{pmatrix} p & q \\ r & s \end{pmatrix}$.

We have $\begin{pmatrix} \psi \\ \varphi \end{pmatrix} = \begin{pmatrix} p & q \\ r & s \end{pmatrix}\begin{pmatrix} u \\ v \end{pmatrix}$ and then $\begin{pmatrix} \psi(z) \\ \varphi(z) \end{pmatrix} = \begin{pmatrix} p\sin\pi z/H \\ r\sin\pi z/H \end{pmatrix}$.

Hence the desired result with $\Psi_0 = p$ and $\Phi_0 = r$.

2. We just have to insert in (10.8) and (10.9).the result obtained in question (1) for $\Psi(z)$ and $\Phi(z)$.
3. This is just relation (10.10). It is the critical equation, since we need it to obtain real eigenmodes.
4. From (10.11) we obtain

$$1 - \frac{k_{\infty,1}}{1+\lambda_0 M_1^2} = \frac{k_{\infty,2}}{(1+\lambda_0 M_1^2)(1+P+\lambda_0 M_2^2)}$$

Which shows that $P > 0$ implies (10.12).

Conversely, if (10.12) holds, if we introduce $P > 0$ as follows in the denominator:

$$\frac{k_{\infty,1}}{1+\lambda_0 M_1^2} + \frac{k_{\infty,2}}{(1+\lambda_0 M_1^2)(1+P+\lambda_0 M_2^2)}$$

By increasing the value of P we decrease the value of this sum of two fractions. It can be made equal to 1 by taking

$$P = \frac{k_{\infty,2}}{1 + \lambda_0 M_1^2 - k_{\infty,1}} - \left(1 + \lambda_0 M_2^2\right)$$

5. We have

$$k_{\infty,2} = \nu p \Sigma_{f2} / \Sigma_2$$

The core is homogeneous. We assume that each cm^3 of the core is made with ω cm^3 of fuel and$(1-\omega)$ cm^3 of moderator, we have (with obvious notations) $\Sigma_{f2} = \omega \Sigma_{fc2}$ and $\Sigma_2 = \omega \Sigma_{ac2} + (1-\omega) \Sigma_{am2}$. Hence

$$\frac{\nu p \Sigma_{f2}}{\Sigma_2} = \frac{\nu p \omega \Sigma_{fc2}}{\omega \Sigma_{ac2} + (1-\omega)\Sigma_{am2}} = \nu p \frac{\Sigma_{fc2}}{\Sigma_{ac2}} \frac{\omega \Sigma_{ac2}}{\omega \Sigma_{ac2} + (1-\omega)\Sigma_{am2}} = p\eta f$$

6. Numerical application: We find $k_{\infty,1} = 0.0484$ and $k_{\infty,2} = 1.022$.
7. We find $P = 0.0699769$.
8. We have $\Psi_0/\Phi_0 = Q\left(1 + P + \lambda_0 M_2^2\right) = 8.272$ so that the fast neutron flux is eight times higher than the thermal neutron flux.
9. From the definition of ε, we should have $\varepsilon = \frac{\nu \Sigma_{f2}\Phi + \nu \Sigma_{f1}\Psi}{\nu \Sigma_{f2}\Phi} = 1 + \frac{\Sigma_{f1}}{\Sigma_{f2}} \frac{\Psi_0}{\Phi_0}$. ■

Solution to Exercise 11.1

1. Deriving (11.7) we find

$$\frac{d^2 n}{dt^2} + \frac{\beta}{\ell}\frac{dn}{dt} = \lambda \frac{dc}{dt} = \lambda \left(\frac{\beta}{\ell} n - \lambda c\right)$$

but

$$\lambda c - \frac{\beta}{\ell} n = \frac{dn}{dt} - S$$

So that

$$\frac{d^2 n}{dt^2} + \frac{\beta}{\ell}\frac{dn}{dt} = \lambda \left(S - \frac{dn}{dt}\right)$$

Hence the result.

2. We know that the general solution of (11.8) is the sum of
   - a particular solution of Eq. (11.8) with RHS
   - the general solution of Eq. (11.8) without RHS that is

$$(*) \quad \frac{d^2 n}{dt^2} + \left(\lambda + \frac{\beta}{\ell}\right)\frac{dn}{dt} = 0$$

Firstly, we easily check that

$$n(t) = \frac{\lambda S}{\lambda + \frac{\beta}{\ell}} t$$

Is a particular solution of (11.8).
Secondly, the general solution of (*) is such that $n(t) = e^{\gamma t}$ and then

$$\gamma^2 + \left(\lambda + \frac{\beta}{\ell}\right)\gamma = 0$$

so that $\gamma = 0$ or $\gamma = -\left(\lambda + \frac{\beta}{\ell}\right)$ hence the result.

In the expression for the general solution $n(t)$ we see that the first two terms are bounded, but the third is not, and so, if $S \neq 0$, $n(t) \to \infty$ when $t \to \infty$.

3. The determinant of matrix A being equal to

$$\frac{1}{\ell}(\lambda(\beta - \rho) - \lambda\beta) = -\lambda\frac{\rho}{\ell}$$

It does not vanish if $\rho \neq 0$, hence the result.

4. We have

$$\frac{\rho - \beta}{\ell} n_1 + \lambda c_1 + S = 0$$

and then (after subtraction)

$$\frac{d}{dt}(n(t) - n_1) = \frac{\rho - \beta}{\ell}(n(t) - n_1) + \lambda(c(t) - c_1)$$

Similarly

$$\frac{\beta}{\ell} n_1 - \lambda c_1 = 0$$

so that

$$\frac{d}{dt}(c(t) - c_1) = \frac{\beta}{\ell}(n(t) - n_1) - \lambda(c(t) - c_1)$$

Hence the result.

5. Let $\alpha = \frac{\rho-\beta}{\ell}$ and $\gamma = \frac{\beta}{\ell}$.

The eigenvalues ω_1 and ω_2 of matrix A are solutions of

$$\begin{vmatrix} \alpha - \omega & \lambda \\ \gamma & -\lambda - \omega \end{vmatrix} = 0$$

That is, they satisfy

$$f(\omega) = \gamma\lambda$$

where

$$f(\omega) = (\omega - \alpha)(\omega + \lambda)$$

To evaluate ω_1 and ω_2, we could use Nordheim's method which would use

$$\omega\ell + \beta\frac{\omega}{\omega + \lambda} = \rho$$

and shows that $\forall\rho$ there exist two real and distinct solutions such that $\omega_1 \leq -\lambda < \omega_2$.

Rather, we shall notice that

$$\omega^2 + (\lambda - \alpha)\omega - (\alpha + \gamma)\lambda = \omega^2 + (\lambda - \alpha)\omega - \frac{\rho}{\ell}\lambda = 0$$

So that

$$\omega_1 \cdot \omega_2 = -\frac{\rho}{\ell}\lambda$$

and then $\omega_2 > 0$ if $\rho > 0$ and $\omega_2 < 0$ if $\rho < 0$.

If $\rho < 0$ both eigenvalues ω_1 and ω_2 are negative; we deduce that $u(t) \to 0$ like $e^{t\omega_2}$. Finally we proved that $u(t) \to 0$ as $t \to \infty$, which is the desired result.

6. We have

$$\frac{\rho - \beta}{\ell} n_1 + \lambda c_1 + S = 0$$
$$\frac{\beta}{\ell} n_1 - \lambda c_1 = 0$$

and then by addition

$$\frac{\rho}{\ell} n_1 + \mathrm{S} = 0$$

We deduce that $\rho = -\frac{\ell}{n_1} S$ which is the desired result. ■

Solution to Exercise 11.2

1. We let $\alpha = \frac{\rho-\beta}{\ell}$ and $\gamma = \frac{\beta}{\ell}$. Note that $\alpha + \gamma = \frac{\rho}{\ell}$.

   Let A be the 2×2 matrix

$$\begin{pmatrix} \alpha & \lambda \\ \gamma & -\lambda \end{pmatrix}$$

   (11.10) and (11.11) is equivalent to

$$\frac{\mathrm{d}}{\mathrm{d}t}\begin{pmatrix} n \\ c \end{pmatrix} = A\begin{pmatrix} n \\ c \end{pmatrix}$$

To determine the asymptotic behavior of solution $\{n(t), c(t)\}$ to this differential system, it is convenient to study the eigenvalues ω_1 and ω_2 of matrix A.

We assume $\omega_1 \leq \omega_2$.

These are solutions of equation:

$$\begin{vmatrix} \alpha - \omega & \lambda \\ \gamma & -\lambda - \omega \end{vmatrix} = 0$$

That is

$$f(\omega) = \gamma\lambda$$

where

$$f(\omega) = (\omega - \alpha)(\omega + \lambda)$$

Equivalently, we have

$$\omega^2 + (\lambda - \alpha)\omega - (\alpha + \gamma)\lambda = \omega^2 + (\lambda - \alpha)\omega - \frac{\rho}{\ell}\lambda = 0 \tag{5}$$

which shows that

$$\omega_1 \cdot \omega_2 = -\frac{\rho}{\ell}\lambda$$

And then that $\omega_2 > 0$ if $\rho > 0$ and that $\omega_2 < 0$ if $\rho < 0$.

When $\rho < 0$, we see that $\{n(t), c(t)\} \to 0$ as t $\to \infty$. *The reactor is therefore subcritical.*

On the contrary, when $\rho > 0$, the product $\omega_1 . \omega_2$ is negative which shows that $\omega_2 > 0$;

Then $\{n(t), c(t)\} \to \infty$ as $t \to \infty$. *The reactor is therefore supercritical.*

As a consequence of both statements, when $\rho = 0$ *the reactor is critical.*

2. When $\rho > \beta$ we have $\alpha > 0$. Let

$$g(\omega) = f(\omega) - \gamma\lambda$$

The eigenvalues ω_1 and ω_2 are solutions of $g(\omega) = 0$.
Note that $f(\alpha) = 0$ so that $g(\alpha) = -\lambda\gamma < 0$;
This shows that $\omega_1 < 0 < \alpha < \omega_2$ so that $n(t) \to \infty$ as $t \to \infty$, faster than $e^{\alpha t}$. In such a case we have prompt-criticality.

3. We now assume that $\rho > 0$ but with $\rho < \beta - \lambda\ell$. We note that $\alpha < -\lambda$.

The curve $\omega \to f(\omega)$ is a parabola going through points $\{\alpha, 0\}$ and $\{-\lambda, 0\}$.
We have $f'(\omega) = 2\omega - \alpha + \lambda$.
For $-\lambda \le \omega$ the curve can be locally assimilated to its tangent in $\{-\lambda, 0\}$, so that

$$f(\omega) \cong f(-\lambda) + f'(-\lambda)(\omega + \lambda) = 0 + (-\lambda - \alpha)(\omega + \lambda)$$

Solving $f(\omega) = \gamma\lambda$ in an approximate way we find

$$\gamma\lambda \cong (-\lambda - \alpha)(\omega_2 + \lambda)$$
$$\omega_2 \cong -\lambda - \frac{\gamma\lambda}{\lambda + \alpha} = -\lambda\frac{\alpha + \lambda + \gamma}{\lambda + \alpha} = -\lambda\frac{\rho + \lambda\ell}{\rho - \beta + \lambda\ell}$$

Since λ is small compared to $\frac{\rho}{\ell}$ and $\frac{\beta-\rho}{\ell}$ then

$$\omega_2 \cong \lambda\frac{\rho}{\beta - \rho}$$

Note that ℓ does not appear in this formula.
We see that if $\rho = \frac{2}{3}\beta$ or $\rho = \frac{3}{4}\beta$, we have $\omega_2 \cong 2\lambda$ or $\omega_2 \cong 3\lambda$. The reactor respond then with a time constant that has the same order of magnitude as the precursor's period If this is of the order of ten seconds or more, the reactor is controllable. The important thing is to keep a sufficient margin with respect to the forbidden value $\rho = \beta$.

4. Numerical application: When $\beta = 380\,\text{pcm}$, $\rho = 190\,\text{pcm}$, $\ell = 2.5\,10^{-5}$ s, $\lambda = 0.09\ \text{s}^{-1}$, we can compute ω_2 exactly by solving the second degree equation $f(\omega) = \gamma\lambda$. We find $\omega_2 = 0.08982996\ \text{s}^{-1}$.

As $\rho = \frac{\beta}{2}$ we may also use the approximate formula that gives $\omega_2 \cong \lambda = 0.09\ \text{s}^{-1}$, which is a very good approximation.

5. From the exact definition of ω_1 and ω_2 we should have $\omega_1 + \omega_2 = \alpha - \lambda$.

Since $\lambda\frac{\rho}{\beta-\rho}$ is a good approximation to ω_2 we can deduce a good approximation to ω_1.

With the numerical data given in question 4, we get $\omega_1 = -76.18$ instead of $\omega_1 = -76.17979$. A very good approximation. Since λ is small compared to α, we even note that α is a good approximation to to ω_1.

6. $\{n(t), c(t)\}$ satisfies

$$\frac{\mathrm{d}n}{\mathrm{d}t} = -\alpha n + \lambda c \quad n(0) = n_0$$
$$\frac{\mathrm{d}c}{\mathrm{d}t} = \frac{\beta}{\ell} n - \lambda c \quad c(0) = c_0$$

Let

$$n(t) = A_1 \mathrm{e}^{\omega_1 t} + A_2 \mathrm{e}^{\omega_2 t}$$
$$c(t) = B_1 \mathrm{e}^{\omega_1 t} + B_2 \mathrm{e}^{\omega_2 t}$$

We have

$$\omega_1 A_1 = -\alpha A_1 + \lambda B_1 \quad \omega_1 B_1 = \frac{\beta}{\ell} A_1 - \lambda B_1 \quad A_1 + A_2 = n_0$$
$$\omega_2 A_2 = -\alpha A_2 + \lambda B_2 \quad \omega_2 B_2 = \frac{\beta}{\ell} A_2 - \lambda B_2 \quad B_1 + B_2 = c_0 = \frac{\beta}{\lambda \ell} n_0$$

We deduce

$$\omega_1 A_1 + \omega_2 A_2 = -\alpha(A_1 + A_2) + \lambda(B_1 + B_2) = -\alpha n_0 + \lambda \frac{\beta}{\lambda \ell} n_0 = \frac{\rho}{\ell} n_0$$
$$\omega_1 B_1 + \omega_2 B_2 = \frac{\beta}{\ell}(A_1 + A_2) - \lambda(B_1 + B_2) = \frac{\beta}{\ell} n_0 - \lambda \frac{\beta}{\lambda \ell} n_0 = 0$$

Solving the 2×2 system satisfied by A_1 and A_2 we get

$$(\omega_2 - \omega_1) A_1 = \left(\omega_2 - \frac{\rho}{\ell}\right) n_0$$
$$(\omega_2 - \omega_1) A_2 = \left(\frac{\rho}{\ell} - \omega_1\right) n_0$$

Similarly for B_1 and B_2

$$(\omega_2 - \omega_1) B_1 = \omega_2 \frac{\beta}{\lambda l} n_0$$
$$(\omega_2 - \omega_1) B_2 = -\omega_1 \frac{\beta}{\lambda \ell} n_0$$

Note that we previously found

$$\omega_2 - \omega_1 = \omega_2 + \omega_1 - 2\omega_1 \cong \alpha - \lambda - 2\alpha = -\alpha - \lambda \cong -\alpha$$

So that

$$-\alpha A_1 \cong \left(-\frac{\rho}{\ell}\right)n_0 \quad \text{hence}\, A_1 = \frac{\rho}{\rho-\beta}n_0$$
$$-\alpha A_2 = \left(\frac{\rho}{\ell}-\alpha\right)n_0 \quad \text{hence}\, A_2 = \frac{\beta}{\beta-\rho}n_0$$
$$-\alpha B_1 = \frac{\rho}{\beta-\rho}\frac{\beta}{\ell}n_0 \quad \text{hence}\, B_1 = \frac{\rho\beta}{(\beta-\rho)^2}n_0$$
$$-\alpha B_2 = -\alpha\frac{\beta}{\lambda l}n_0 \quad \text{hence}\, B_2 = \frac{\beta}{\lambda l}n_0$$

We see that the solution will move rapidly from n_0 to $n(t) = A_2 = \frac{\beta}{\beta-\rho}n_0$ and then increase slowly like $\frac{\beta}{\beta-\rho}n_0\exp\left(\lambda\frac{\rho}{\beta-\rho}t\right)$. The rapid evolution from n_0 to $\frac{\beta}{\beta-\rho}n_0$ is a prompt effect. See figure below, where we can see that, with the above data, the power is rapidly multiplied by 2.

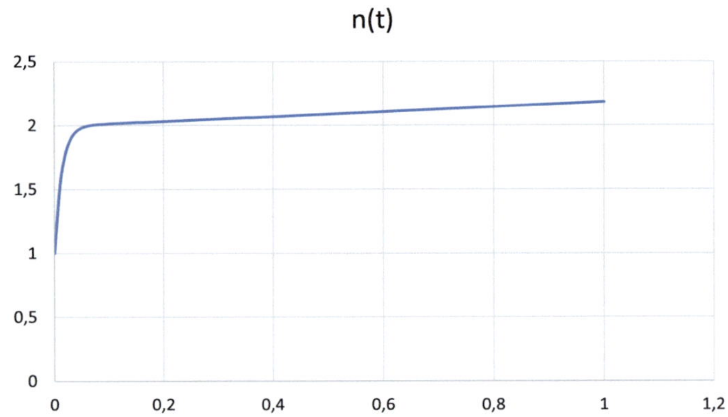

■

Solution to Exercise 12.1

1. Let $J(t) = e^{\lambda t}I(t)$.

   We have $\frac{d}{dt}J(t) = e^{\lambda t}\left(\frac{d}{dt}I(t)+\lambda I\right) = e^{\lambda t}f$.
   Then:

$$J(t) = J(0) + \int_0^t \frac{d}{ds}J\,ds = J(0) + \int_0^t e^{\lambda s}f\,ds = J(0) + f\left[\frac{e^{\lambda s}}{\lambda}\right]_0^t$$
$$= J(0) + f\frac{e^{\lambda t}-1}{\lambda}$$

We deduce

$$I(t) = e^{-\lambda t} J(t) = I(0)e^{-\lambda t} + f\frac{1 - e^{-\lambda t}}{\lambda}$$

Numerical application: We find

$$f = 5.856\,10^{11}\ \text{at/cm}^3/\text{s}$$

and then

$$I_\infty = 2.05\,10^{16}\ \text{at/cm}^3$$

2. Let $Y(t) = e^{\mu t} X(t)$.

   We have $\frac{d}{dt}Y(t) = e^{\mu t}\big(\frac{d}{dt}X(t) + \mu X\big) = e^{\mu t}(g + \lambda I) = e^{\mu t}\big(g + f\big(1 - e^{-\lambda t}\big)\big)$.
   Then, noting that $Y(0) = 0$,

$$\begin{aligned} Y(t) &= \int_0^t \frac{d}{ds} Y\,ds = \int_0^t e^{\mu s}\big(g + f\big(1 - e^{-\lambda s}\big)\big)ds \\ &= \int_0^t (f + g)e^{\mu s}ds - f\int_0^t e^{(\mu-\lambda)s}ds = (f + g)\frac{e^{\mu t} - 1}{\mu} - f\frac{e^{(\mu-\lambda)t} - 1}{\mu - \lambda} \end{aligned}$$

   Coming back to $X(t) = e^{-\mu t} Y(t)$, we get

$$X(t) = (f + g)\frac{1 - e^{-\mu t}}{\mu} - f\frac{e^{-\lambda t} - e^{-\mu t}}{\mu - \lambda}$$

   We deduce $X_\infty = \frac{1}{\mu}(f + g)$. ■

Numerical application: We find

$$\begin{aligned} g &= 9.15\,10^{9}\ \text{at/cm}^3/\text{s} \\ \mu &= 2.08\,10^{-5}\ \text{s}^{-1} \\ X_\infty &= 6.33\,10^{15}\ \text{at/cm}^3 \end{aligned}$$

3. The first result is straightforward. To prove the second one, we let $Y(t) = e^{\mu t}X(t)$.

   We have

$$\frac{d}{dt}Y(t) = e^{\mu t}\left(\frac{d}{dt}X(t) + \mu X\right) = e^{\mu t}(\lambda I) = e^{\mu t}\big(\lambda I(0)e^{-\lambda t}\big).$$

$$Y(t) = Y(0) + \int_0^t \frac{\mathrm{d}}{\mathrm{d}s} Y \mathrm{d}s = Y(0) + \lambda I(0) \int_0^t \mathrm{e}^{(\mu-\lambda)s} \mathrm{d}s$$
$$= Y(0) + \lambda I(0) \frac{\mathrm{e}^{(\mu-\lambda)t} - 1}{\mu - \lambda}$$
$$X(t) = \mathrm{e}^{-\mu t} Y(t) = \mathrm{e}^{-\mu t} X(0) + \lambda I(0) \frac{\mathrm{e}^{-\lambda t} - \mathrm{e}^{-\mu t}}{\mu - \lambda}. \quad \blacksquare$$

4. We have:

$$X(t) = \mathrm{e}^{-\mu t} X(0) - A\left(\mathrm{e}^{-\lambda t} - \mathrm{e}^{-\mu}\right) = (X(0) + A)\mathrm{e}^{-\mu t} - A\mathrm{e}^{-\lambda t}$$

Thus

$$\frac{\mathrm{d}}{\mathrm{d}t} X(t) = -\mu(X(0) + A)\mathrm{e}^{-\mu t} + \lambda A \mathrm{e}^{-\lambda t}$$

If $X(t)$ has a maximum, then $\frac{\mathrm{d}}{\mathrm{d}t} X = 0$ and t is such that:

$$\mu(X(0) + A)\mathrm{e}^{-\mu t} = \lambda A \mathrm{e}^{-\lambda t}$$
$$\mathrm{e}^{(\lambda-\mu)t} = \frac{\lambda A}{\mu(x(0) + A)}$$

hence $t = \frac{1}{\lambda - \mu} \ln \frac{\lambda A}{\mu X(0) + \mu A}$.

Numerical application: We find:

$$A = 7.51\ 10^{16}\ \text{at/cm}^3$$
$$t = 30{,}500\ \text{s i.e. } 8.46\ \text{h}. \quad \blacksquare$$

Solution to Exercise 12.2

1. At equilibrium we have $\frac{\mathrm{d}P}{\mathrm{d}t} = \frac{\mathrm{d}S}{\mathrm{d}t} = 0$ so that, in the fuel:

$$S = \frac{\gamma \Sigma_f}{\sigma} = 7.04\ 10^{16} \quad \text{at/cm}^3 \left(\text{and } P = \gamma \Sigma_f \Phi / \lambda\right)$$

The impact of Samarium is to increase from Σ_a to $\Sigma_a + \omega\sigma S = \Sigma_a + \omega\gamma\Sigma_f$ the core macroscopic absorption cross section.

On the other hand, the core macroscopic fission cross section remains equal to Σ_f; the infinite multiplication factor k_∞ decreases then from

$$k_\infty = \nu_e \frac{\omega \Sigma_f}{\Sigma_a} \quad \text{to} \quad k_\infty = \nu_e \frac{\omega \Sigma_f}{\Sigma_a + \omega\sigma S} = \nu_e \frac{\omega \Sigma_f}{\Sigma_a + \omega\gamma\Sigma_f}$$

So that it is divided by

$$\frac{\Sigma_a + \omega\gamma\Sigma_f}{\Sigma_a} = 1 + \omega\frac{\gamma\Sigma_f}{\Sigma_a}. \left(\text{or by } \frac{\Sigma_a + \omega\sigma S}{\Sigma_a} = 1 + \omega\frac{\sigma S}{\Sigma_a}.\right)$$

Samarium brings then a negative reactivity equal to $\omega\frac{\gamma\Sigma_f}{\Sigma_a}\,10^5$ (in pcm).

2. Numerical application: We find -553 pcm.
3. We find $S = 7.04\,10^{16}$ at/cm^3.
4. We find $P = 5.04\,10^{16}$ at/cm^3.
5. When $\Phi = 0$, the first equation shows that $P(t)$ will slowly decrease.

The second equation shows that $S(t)$ will exceed the value calculated in question 3. So, on restart, the negative reactivity will be higher than − 553 pcm.

More precisely, we have

$$\frac{dS}{dt} = \lambda P = -\frac{dP}{dt}$$

Thus (taking reactor shutdown as time origin):

$S(\infty) - S(0) = P(0) = \frac{1}{\lambda}\gamma\Sigma_f\Phi$ (where Φ was the neutron flux before reactor shutdown).

The number of Samarium atoms will then increase from $S(0) = 7.04\,10^{16}$ at/cm^3 to

$$S(\infty) = S(0) + P(0) = 12.08\,10^{16}\ \text{at/cm}^3$$

which corresponds to a − 948 pcm poisoning.

Then, following reactor restart, this negative reactivity will decrease to − 553 pcm after dozens of hours. ■

Solution to Exercise 12.3

1. First, we note that $\Sigma_f^U\Phi$ is the number of fissions per fuel cm^3 per second; $\gamma\Sigma_f\Phi$ is then the number of $^{103}_{45}$Rh atoms produced per fuel cm^3 per second.

On the other hand, σR being the macroscopic capture cross section of $^{103}_{45}$Rh, $\sigma\Phi R$ is the number of $^{103}_{45}$Rh atoms that are transformed into $^{104}_{45}$Rh in each fuel cm^3 each second, the RHS of (12.3) is therefore the difference between production and loss.

2. To solve (12.3) we first look for the general solution of

$$\frac{dR}{dt} + \sigma\Phi R = 0$$

Then we add a particular solution of (12.3) (we choose $R(t) = \gamma \Sigma_f^U / \sigma$), thus we find:

$$R(t) = A e^{-\sigma \Phi t} + \gamma \Sigma_f^U / \sigma$$

Since $R(0) = 0$, we deduce $A = -\gamma \Sigma_f^U / \sigma$ hence $R(t) = \left(1 - e^{-\sigma \Phi t}\right) \gamma \Sigma_f^U / \sigma$, which gives the desired result.

3. Proceeding as for xenon poisoning we see that the Rhodium poisoning must be equal to $10^5 \frac{\omega \sigma R}{\Sigma_0} = 10^5 \frac{\omega \gamma_f^U}{\Sigma_0}$ pcm that is 3072 pcm. *It is interesting to note that this poisoning is independent of the value of* σ!

However, the speed at which such poisoning is achieved depends on σ: in a PWR we need 1000 *days* to get $\left(1 - e^{-\sigma t}\right) > 0.82$. ■

Solution to Exercise 13.1

1. On the RHS of (13.1) $\Sigma_{c8}\ \Phi$ is the number of ^{238}U capture per cm^3 per second, then it is the production of ^{239}Np. The LHS is the differential equation describing the evolution of ^{239}Np. Then λ should be selected as the ^{239}Np β-decay constant. More precisely $\lambda = \ln(2)/T$ where $T = 55.2\,h = 198720\,s$. Then $\lambda = 3.488\,10^{-6}\,s^{-1}$.

Similarly the RHS of (13.2) is the production of ^{239}Pu by ^{239}Np β-decay. Since the radioactive period of ^{239}Pu is very long (24,000 years) the second term of the LHS should be equal to the absorption of neutrons by ^{239}Pu. More precisely, we must choose $\mu = \sigma_{a9}\Phi$. Then $\mu = 3.66\,10^{-8}$ n/s.

2. By solving (13.1), we have

$$M_9(t) = \frac{1}{\lambda}\Sigma_{c8}\left(1 - e^{-\lambda t}\right)\Phi$$

Reporting in (13.2) we have

$$\frac{d}{dt}N_9 + \mu N_9 = \Sigma_{c8}\left(1 - e^{-\lambda t}\right)\Phi$$

Let $P_9(t) = e^{\mu t} N_9(t)$ we have

$$\frac{d}{dt}P_9 = e^{\mu t}\left(\frac{d}{dt}N_9 + \mu N_9\right) = e^{\mu t}\left(\Sigma_{c8}\left(1 - e^{-\lambda t}\right)\Phi\right)$$

Hence $P_9(t) = C + \Sigma_{c8}\Phi\left(\frac{e^{\mu t}}{\mu} - \frac{e^{(\mu-\lambda)t}}{\mu-\lambda}\right)$ where C denotes some constant.

We should have $C = -\left(\Sigma_{c8}\Phi\left(\frac{1}{\mu} - \frac{1}{\mu-\lambda}\right)\right.$ since $N_9(0) = 0$. Finally:

$$N_9(t) = -\Sigma_{c8}\Phi\left(\frac{1}{\mu} - \frac{1}{\mu-\lambda}\right)\mathrm{e}^{-\mu t} + \Sigma_{c8}\Phi\left(\frac{1}{\mu} - \frac{\mathrm{e}^{-\lambda t}}{\mu-\lambda}\right)$$
$$= \Sigma_{c8}\Phi\left[\frac{1}{\mu}\left(1-\mathrm{e}^{-\mu t}\right) - \frac{1}{\mu-\lambda}\left(\mathrm{e}^{-\lambda t} - \mathrm{e}^{-\mu t}\right)\right]$$

So we should have $\Sigma_{c8}\Phi = \lambda M_9^\infty = \mu N_9^\infty$, which is a quite intuitive result. In particular $N_9^\infty = \frac{\Sigma_{cs}\Phi}{\sigma_{a9}\Phi} = \frac{\Sigma_{cs}}{\sigma_{a9}} = \frac{\sigma_{cs}}{\sigma_{a9}}N_8 = \frac{2.7}{1000}N_8$.

3. When μt is big we have $N_9 \cong 0.0027 N_8$.

   On the other hand $\mu N_9 \cong \Sigma_{a9}\Phi \cong \Sigma_{c8}\Phi$.
   Therefore $\Sigma_{c9}\Phi + \Sigma_{f9}\Phi \cong \Sigma_{c8}\Phi$ in steady state.
   The physical interpretation is as follows.

- there are as many U8 nuclei that produce Pu9 as Pu9 nuclei that disappear by capture or fission
- Uranium fuel permanently contains 0.27% Pu9.

4. We notice that μ is 100 times smaller than λ. Then the time evolution of N_9 will be 100 times longer than for M_9.

   More precisely, when t is larger than 10 days $\mathrm{e}^{-\lambda t}$ is negligible, so that

$$N_9(t) \cong \Sigma_{c8}\Phi\left(\frac{1}{\mu} - \left(\frac{1}{\mu} - \frac{1}{\mu-\lambda}\right)\mathrm{e}^{-\mu t}\right) = \Sigma_{c8}\Phi\left(\frac{1}{\mu} + \frac{\lambda}{\mu(\mu-\lambda)}\mathrm{e}^{-\mu t}\right)$$
$$\cong \frac{1}{\mu}\Sigma_{c8}\Phi\left(1-\mathrm{e}^{-\mu t}\right)$$

   The latter is clearly the solution of (13.3).

Solution to Exercise 13.2

1. Each FPED, we must produce 86,400 × 2.7 GJ therefore the fissioned mass should be equal to 2844 g for U5 + Pu + U8.

   If we except U8 the fissioned mass should be equal to 2841 g/FPED.
   Each time we consume 1 g $U5$ we fission 0.85 g $U5$ and produce $0.6 \times 0.73 = 0.438$ g Pu.
   Therefore we must have $\delta m \times (0.85 + 0.438) = 2841$ that is $\delta m = 2841/1.288 = 2206$ g/FPED.

2. The initial U5 mass is $m = 3184$ kg.

   We must have $x = \frac{m - m_0}{\delta m} = \frac{3184-1468}{2.2} = 778$ FPED.
   The residual enrichment is $\frac{1468}{79{,}600} = 1.8\%$.
   The energy produced is $2.7 \times 778 = 2100$ GWd.
   The burn-up is $BU_1 = 2100/79.6 = 26.4$ GW$d/t$.

3. At BOC, the U5 mass in the new half-core is $m/2$. The U5 mass in the other half core that already been irradiated for one cycle is $(m - x\,\delta m)/2$. At EOC, the U5 masses became $(m - x\,\delta m)/2$ et $(m - 2x\delta m)/2$ respectively. To maximize x we should have $m - 3x\delta m/2 = m_0$ then

$$x = 2(m - m_0)/3\delta m = 778 \times 2/3 = 518\,\text{FPED}.$$

At EOC, the most used half core contains 1592 − 518 * 1.9 = 607 kg U5.
Its residual enrichment is then 607/39,800 = 1.53%.
In 1036 FPED the most used half core will have produced 2.7 * 518 = 1399 GWd.
Which corresponds to a 35 GW$d/t$ burn-up.

4. With such a management, sub-part $n°i$ contains U5

$$m_i = (m - ix\delta m)/\text{n} \;\; \text{kg}$$

at EOC for $i = 1, \ldots, n$ (it contained $(m - (i - 1)\,x\,\delta m)/n$ at BOC).

The longest possible cycle is such that

$$m_1 + m_2 + \cdots + m_0 = m_0$$

Now, $m_1 + m_2 + \cdots + m_n = m - (1 + 2 + \cdots + n)\,x\,\delta m/n$.
Noting that $1 + 2 + \ldots + n = n(n + 1)/2$, we must have:

$$m - (n + 1)x\delta m/2 = m_0$$

hence $x = 2/(n + 1).\,(m - m_0)/\delta m$.
With the provided data we have $(m - m_0)/\delta m = 778$.
So, we find $x = 518$ FPED when $n = 2$, that we found at question 3.
We find $x = 389$ FPED for $n = 3$ and $x = 311$ FPED for $n = 4$.
For $i = n$ we find that the residual U5 mass in the most irradiated sub-part is:

$$m_n = (m - nx\delta m)/n = m/n - 2/(n + 1) \cdot (m - m_a)$$

For $n = 4$ we find

$$m_n = 796 - 2 * 1474/5 = 206\,\text{kg}$$

Its residual enrichissement is then 206/19,900 = 1.03%.

5. The U5 mass consumed on sub-part $i = n$ is

$$m/n - m_n = 2/(n + 1) \cdot (m - m_0) = 2n/(n + 1) \cdot (m - m_0)/n$$

hence the result since the energy produced is proportional to the U5 mass consumed, and $(m - m_0)/n$ is effectively the U5 mass that would be consumed on sub-part $i = n$ in the managing mode where all fuel assemblies would be removed at the same time.

The burn-up will be $BU_n = 2n/(n + 1).\ BU_1$.

n	$\mathrm{BU}_n(\mathrm{GWd}/t)$
1	26.4
2	35.2
3	39.6
4	42.24

Note that this table explains Fig. 13.4. Indeed, with a given enrichment (here 4%) the higher n the higher the Burn Up. The counterpart is that the higher n, the shorter the cycle duration.

Solution to Exercise 13.3

1. We see that 955–910 = 45 kg ^{238}U have disappeared; 39 kg by capture.
2. We see that 39 kg ^{239}Pu have been created but 5 kg only remained at the end, so 34 kg disappeared: 3/4 by fission (25.5 kg) and 1/4 by capture (8.5 kg).
3. From the previous question we see that ^{240}Pu = 8.5 kg have been created; since 2.4 kg ^{240}Pu remain, we deduce that ^{241}Pu = 8.5–2.4 = 6.1 kg have been created.
4. The ^{241}Pu mass that has been fissioned is 6.1–2.2–0.94 = 2.96 kg.
5. The fissioned ^{235}U mass is 45-9 -5.49 = 30.51 kg.
6. The mass of fissioned nuclei is

$$6\left(\text{for}\,^{238}\text{U}\right) + 30.51\left(^{235}\text{U}\right) + 25.5\left(^{239}\text{Pu}\right) + 2.96\left(^{241}\text{Pu}\right) = 64.97\,\text{kg}$$

7. Since fission produces 82 GJ/g the energy produced is 64.97 $\times$ 82 = 5327 TJ which represents 62 GWd. So, since initially there was 1t of heavy metal, the burn up is 62 GWd/t.

Solution to Exercise 13.4

1. The table given in the text of the exercise gives both for ^{235}U and ^{238}U the number of captures and the number of fissions per cm^3 per second for the 4 groups. By summation, we can evaluate that the ^{235}U consumption is 1.49 10^{13} at/cm^3/s. The number of ^{238}U captures is 8.99 10^{12} c/cm^3/s. The conversion factor is then $C =$ 0.603.
2. Let F_5 denote the number of neutrons generated by ^{235}U fission: we have

$$F_5 = 2.42 \times 1.04\,10^{13} = 2.51\,10^{13}\,\text{n/cm}^3/\text{s}.$$

The number of extra neutrons generated by fast fisson of ^{238}U is

$$F_8 = (2.9 - 1) \times 6.62\,10^{11} = 1.26\,10^{12}\ \text{n/cm}^3/\text{s}.$$

Then we have

$$\varepsilon = \frac{F_5 + F_8}{F_5} = 1.05$$

To evaluate p we compute $F_5 + F_8 = 2.64\,10^{13}$ that is the number of fast neutrons entering slowing down.

Then we evaluate the number of neutron captures during slowing down

$$\text{NCSD} = 6.67\,10^{12} + 1.41\,10^{12} + 4.99\,10^{10} = 8.13\,10^{12}\ \text{c/cm}^3/\text{s};$$

Then we have $p = \frac{F_5+F_8-\text{NCFSD}}{F_5+F_8} = 0.69$.

For η we evaluate first the fission probability in Gr 4: $P_f = \frac{8.68\,10^{12}}{(8.68+2.34+3.08)\,10^{12}} = 0.616$: we deduce $\eta = \nu.P_f = 1.49$.

3. The total number of absorptions in Gr. 4 (for the fuel) is

$$A_5 + A_8 = 1.17\,10^{13} + 2.34\,10^{12} = 1.41\,10^{13}\ \text{a/cm}^3/\text{s}$$

For the moderator it is $\Sigma_c^H \cdot \Phi_4 = 3.66\,10^{11}$ c/cm^3/s.

For the core it is $\frac{2\Sigma_c^H \cdot \Phi_4 + A_5 + A_3}{3} = 4.39\,10^{12}$ a/cm^3/s.

Then $f = \frac{A_5+A_8}{2\Sigma_C^H \cdot \Phi_4 + A_5 + A_8} = 0.95$.

Finally we deduce $k_\infty = 1.03$. ■

Solution to Exercise 13.5

We apply Fig. 13.6 with $C = {}^{233}$Pa, $D = {}^{233}$U, $E = {}^{234}$Pa, and $B = {}^{232}$Th (doing so, we neglect the ^{233}Th intermediate).

We have

$$\frac{\mathrm{d}}{\mathrm{d}t} N_C = \sigma_{cB} N_B \Phi - \lambda_C N_C - \sigma_{aC} N_C \Phi$$

In steady state, we have $\frac{\mathrm{d}}{\mathrm{d}t} N_C = 0$ and then.

$$\sigma_{cB} N_B \Phi = \lambda_C N_C + \sigma_{aC} N_C \Phi$$

On the LHS we have a production term; on the RHS a disappearance term that is composed of

- ^{233}U production
- ^{234}Pa production.

For both terms to be equal we must have $\lambda_C N_C = \sigma_{aC} N_C \Phi$ hence

$$\Phi = \frac{\lambda_C}{\sigma_{aC}} = \frac{2.93\ 10^{-7}}{4.3\ 10^{-23}} = 6.8\ 10^{15}\ \text{n/cm}^2/\text{s}$$

When the neutron flux is smaller ^{233}Pa will produce more ^{233}U than ^{234}Pa.

Ultimately, if we take ^{233}Pa out of the core (which is possible with molten salt reactors) the ^{233}U production will be maximized. ■

Solution to Exercise 14.1

1. We should take $\Sigma_0 = \omega \Sigma_a^U + (1-\omega)\Sigma_c^H$ and $\Sigma_f = \omega \Sigma_f^U$ where Σ_a^U (resp. Σ_f^U.) is the UO_2 macroscopic absorption (resp. fission) cross section, and Σ_c^H the light water macroscopic capture cross section.
2. We have

$$\Sigma_0 = \omega \Sigma_c^U + (1-\omega)\Sigma_c^H = 0.2 * 0.681 + 0.8 * 0.0215 = 0.153\,\text{cm}^{-1}$$
$$\Sigma_f = 0.1054\,\text{cm}^{-1}$$

3. From the definition of the dilution cross section, we have $\sigma_d = \frac{(1-\omega)N_H}{\omega N_U}\sigma_{sH}$.

Taking $N_H = 0.067\ 10^{24}$ at/cm^3 and $N_U = 0.0245\ 10^{24}$ at/cm^3 as in Exercise 7.3, we find $\sigma_d = 218\,b$.

For 3.7% enriched UO_2, we find on Fig. 7.1 $I_{\text{eff}} \cong 47b$ and then

$$p = \exp\left(-\frac{47}{218}\right) = 0.806$$

so that $\nu_e = \nu\,.\,\varepsilon\,.\,p = 2.11$.

We deduce $k_\infty = \nu_e \cdot \frac{\Sigma_f}{\Sigma_0} = 1.454$ (and also $f = 0.887$ and $\eta = 1.934$).

4. From the critical equation, we must have

$$\lambda_0 M^2 = k_\infty - 1 = 0.454$$

hence $\lambda_0 = \frac{0.454}{50} = 0.0908\,\text{cm}^{-2} = \left(\frac{\pi}{R}\right)^2$.

The critical radius is then

$$R = \frac{\pi}{\sqrt{\lambda_0}} = 33\text{cm}$$

so that the critical volume is a sphere the volume of which is about 150 dm^3.

It contains 30 dm^3 of UO_2 that is 330 kg.

We deduce that it contains about $m_1 = 290$ kg of Uranium (10.7 kg of ^{235}U).

5. λ_0 is a solution to a second-degree equation which, insofar as $k_\infty > 1$, has a positive and a negative root. With $k_\infty = 1.454$ The positive root is $\lambda_0 = 0.00873$ cm^{-2}.

   We deduce $R = 33.61$ cm so that $m_2 = 308$ kg.

This is not very far from the result obtained with 1 group, insofar as the sum of the two migration areas (rapid + thermal) is equal to 50 cm^2, like the 50 cm^2 chosen for the 1-group model.

However, as we already noticed in remark 10.1, the choice of the two migration areas is fairly natural in the 2-groups model, whereas it is completely artificial in the 1-group model.

The proper choice of migration area parameter(s) is completely essential for these simplified models. ■

Solution to Exercise 14.2

1. We find $N_{\text{PuO}_2} = 0.02556\,10^{24}$ molecules/cm^3 and then $\Sigma_a = 0.05009\,\text{cm}^{-1}$, $\Sigma_f = 0.04856\,\text{cm}^{-1}$, $\Sigma_{s\text{Pu}} = 0.2556\,\text{cm}^{-1}$, $\Sigma_{sO} = 0.2045\,\text{cm}^{-1}$ and $\Sigma_s = 0.4601\,\text{cm}^{-1}$.
2. From the answer to the previous question we find $\Sigma_t = \Sigma_s + \Sigma_a = 0.5101\,\text{cm}^{-1}$. The mean free path, which is $1/\Sigma_t$ is then about 2 cm.

   Probability of (a) is $\Sigma_{sO}/\Sigma_t = 40\%$, probability of (b) is $\Sigma_{sPu}/\Sigma_t = 50\%$, probability of (c) is $\Sigma_{aPu}/\Sigma_t = 10\%$.
3. Only elastic collisions on an oxygen atom are likely to slow down neutrons. A simple calculation shows that the slowing down is weak since the lethargy of the neutron increases only by 0.12. It would therefore take at least 100 collisions of this type for the neutron to become thermal. However, on average after 4 collisions of this type, the neutron will be absorbed and with a high probability produce a fission that will generate 2.91 fast neutrons. So all the neutrons that migrate into the ball are fast.
4. Obviously we must have $\Phi = \Phi_0$ (up to a multiplicative constant) and $\lambda = \big(\nu\Sigma_f - \Sigma_a\big)/D$ hence the desired result.
5. We have $k_\infty = \nu\sigma_f^9/\sigma_a^9$, We find $k_\infty = 2.82$.
6. Let $\Sigma_t = \Sigma_s + \Sigma_a$ we compute D by using the formula $D = \frac{1}{3\Sigma_t}$, and then M^2.

   We find $D = 0.653\,\text{cm}$ then $M^2 = 13.04\,\text{cm}^2$.
7. We find $R = 8.408$ cm. We should logically deduce that the critical mass is 28.6 kg.
8. the true boundary condition is not (14.3) but:

$$J_- = \frac{1}{4}\Phi_0 + \frac{D}{2}\frac{\partial}{\partial r}\Phi_0 = 0$$

   which means that the input neutron current in V is zero.

Since D is small compared to R, we notice that

$$\Phi_0(R+2D) \cong \Phi_0(R) + 2D\frac{\partial}{\partial r}\Phi_0(R) = 0$$

So that the true Φ_0 does vanish at $r = R+2D$. So the critical sphere has for radius $8.408 - 2 \times 0.653 = 7.1$ cm.

Our method then gives a critical mass of 17.2 kg which is closer to the true value that is 10 kg.

Now the true value is obtained with a Monte-Carlo code which.

- Solves the transport equation (and not the diffusion equation)
- uses a fine description of the dependence in energy of the microscopic cross-sections which we cannot have with a one-group model. ■

Solution to Exercise 15.1

1. Assume $V \complement \mathbb{R}^3$ is an arbitrary domain with boundary Γ sufficiently regular, energy conservation requires that, (if $\vec{n} = \vec{n}(x)$ denotes the unit normal vector to V in $x \in \Gamma$, outwards directed)

$$\int_V P\mathrm{d}x = \int_\Gamma \vec{F}.\vec{n}\mathrm{d}\Gamma$$

Indeed, the LHS is the global power released in V by P and the RHS is the net balance of outgoing and ingoing heat through Γ.

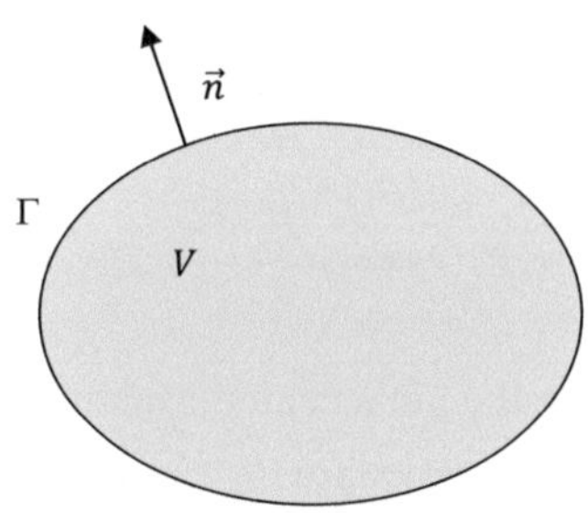

The divergence theorem (cf. Wikipedia) states that for a differentiable vector field. $\vec{F} = \vec{F}(x)$, we have

$$\int_\Gamma \vec{F}.\vec{n}\mathrm{d}\Gamma = \int_V \mathrm{div}\,\vec{F}\mathrm{d}x \tag{1}$$

We deduce

$$\int_V P\mathrm{d}x = \int_V \mathrm{div}\,\vec{F}\mathrm{d}x$$

and since V is arbitrary, we must have

$$\mathrm{div}\vec{F} = P$$

2. In *cylindrical* coordinates, we have $\overrightarrow{\nabla \mathrm{T}} = \left(\frac{\partial T}{\partial r}, \frac{1}{r}\frac{\partial T}{\partial \theta}, \frac{\partial T}{\partial z}\right)$.

   In case of cylindrical symmetry, T depends on neither θ nor z, hence the result.
3. Obvious.
4. From the definition of F_0 we have $\frac{1}{r}\frac{\mathrm{d}}{\mathrm{d}r}(rF_0) = P$ so that $\frac{\mathrm{d}}{\mathrm{d}r}(rF_0) = Pr$.

   Solving this differential equation we have

$$rF_0 = \frac{P.r^2}{2} + k \text{ hence } F_0 = \frac{P.r}{2} + \frac{k}{r}$$

Physically the thermal flux vanishes at the pin center.

Mathematically, we introduce a small cylinder V with radius ε and height L around pin's axis.

The volume of such a cylinder is equal to $L\pi\varepsilon^2$. The global power released in V is then $PL\pi\varepsilon^2$.

Now, let us compute the thermal flux outgoing through the lateral surface of V (whose surface is a $2\pi\varepsilon L$).

$$2\pi\varepsilon L F_0(\varepsilon) = 2\pi\varepsilon L\left(\frac{P.\varepsilon}{2} + \frac{k}{\varepsilon}\right) = PL\pi\varepsilon^2 + 2\pi k L$$

We deduce that $k = 0$, hence $F_0(r) = \frac{P.r}{2}$.

Then we have $\frac{\mathrm{d}}{\mathrm{d}r}T_0 = -\frac{P.r}{2\lambda_0}$ which implies

$$T_0(r) = -\frac{P.r^2}{4\lambda_0} + T_c$$

where T_c is the temperature at the pin's center. Making $r = a$ we get $T_a = T_0(a) = T_c - \frac{P.a^2}{4\lambda_0}$ hence $T_c = T_a + \frac{P.a^2}{4\lambda_0}$; we deduce $T_0(r) = T_a + \frac{P(a^2-r^2)}{4\lambda_0}$ that is the desired result.

5. From the definition of F_1 we have $\frac{1}{r}\frac{\mathrm{d}}{\mathrm{d}r}(rF_1) = 0$ and then $\frac{\mathrm{d}}{\mathrm{d}r}(rF_1) = 0$ so that

$$rF_1 = c \text{ hence } F_1 = \frac{c}{r}$$

We deduce that

$$-\lambda_1 T_1 = c \ln r + d$$

which is equivalent to the desired result. We now write that $T_1(b) = T_b$, which gives

$$d = -\lambda_1 T_b - c \ln b, \quad \text{and}$$

$$T_1(r) = -\frac{1}{\lambda_1}(c \ln r - \lambda_1 T_b - c \ln b) = T_b + \frac{c}{\lambda_1} \ln \frac{b}{r}$$

6. Physically, the temperature may have a slope discontinuity at the fuel/cladding interface, but the thermal flux must be continuous, so that

$$F_0(a) = F_1(a)$$

We obtain $\frac{P.a}{2} = \frac{c}{a}$ so that $c = \frac{P.a^2}{2}$.

7. Since there is no contact thermal resistance at the fuel pellet/cladding interface, we shall have $T_0(a) = T_1(a)$, i.e. temperature continuity between fuel and cladding.

Now $T_1(a) = T_a = T_b + \frac{c}{\lambda_1} \ln \frac{b}{r} = \frac{P.a^2}{2\lambda_1} \ln \frac{b}{a} + T_b$.
Hence $T_0(r) = T_a + \frac{P(a^2 - r^2)}{4\lambda_0} = \frac{P(a^2-r^2)}{4\lambda_0} + \frac{P.a^2}{2\lambda_1} Ln \frac{b}{a} + T_b$.

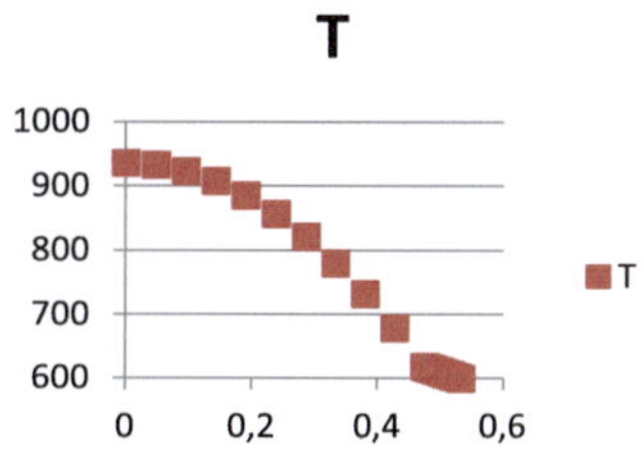

Numerical application: we find $T_a = 618$ K and $T_c = 936$ K. On the curve above, we observe a slope discontimuity in $r = a$.

It comes from the fact that the flux is continuous at the interface $r = a$, but the conductivity is not. ■

References

1. G. Bell, S. Glasstone, *Nuclear Reactor Theory* (Van Nostrand Reinhold, 1970)
2. P. Reuss, *Neutron Physics* (EDP-Sciences, 2008)
3. E. E. Lewis, *Fundamentals of Nuclear Reactor Physics* (Academic Press, 2008)
4. J. Guidez et al., *Les réacteurs nucléaires à caloporteur sodium* (CEA, Monographie DEN, 2014)
5. B. Barré et al., *Nuclear Reactor Systems*, vol 91 (EDP Sciences, Les Ulis, France)
6. DOE Fundamentals Handbook. in *Nuclear Physics and Reactor Theory* (1993)
7. B. Mercier, P. Jinghan, Solving the neutron slowing down equation and computing self-shielded cross sections, in *Proceedings of ICAPP 2015*. Nice (France), Paper 15543, 03–06 May 2015
8. R. Dautray, J.L. Lions, *Mathematical Analysis and Numerical Methods for Science and Technology* (Springer, 1999)
9. S. Marguet, *La Physique des réacteurs nucléaires* (Lavoisier, 2011)
10. J.-L. Nigon, *CNAM textbook* (2009, unpublished)
11. P. Cousinou, *IRSN, introduction to criticality risk* (1995)
12. J. Collinet, *Introduction à la sûreté nucléaire* (Cours IFCEN, 2015, unpublished)
13. H. Grard, *Physique Fonctionnement et Sûreté des REP. Le réacteur en fonctionnement* (EDP Sciences, 2014)
14. P. Reuss, B. Mercier, A simple model to show the effect of counter-reactions. EPJ Nucl. Sci. Technol. **9**, 16 (2023)
15. : B. Tarride, *Physique, fonctionnement et sûreté des REP—Maitrise des situations accidentelles du système réacteur* (EDP-Sciences, 2013)
16. https://www-pub.iaea.org/MTCD/Publications/PDF/TCS-22_2nd_web.pdf
17. B. Mercier, B. Tarride, *Un modèle simple du PWR—Archive ouverte HAL.* https://hal.science/hal-04816281
18. B. Mercier, Z. Ziliang, C. Liyi, S. Nuoya, *Modeling and control of xenon oscillations in thermal neutron reactors.* EPJ Nucl. Sci. Technol. **6**, 48 (2020). 10.1051/epjn/2020009
19. G. Mathonnière, *Etude de problèmes neutroniques liés à la présence du Xénon dans les réacteurs à eau pressurisée.* Thèse (1988)
20. A. Santamarina, *Private communication* (2019)
21. G.S. Lellouche, Space dependent Xenon oscillations. Nucl. Sci. Eng. **12**(1962), 482–489
22. Y. Shimazu, Continuous guidance procedure for Xenon oscillation control. J. Nucl. Sci. Technol. **32**(11), 1159

B. Mercier, *Simple Models for Understanding Nuclear Reactor Physics*,
https://doi.org/10.1007/978-3-031-98532-4

Index

B. Mercier, *Simple Models for Understanding Nuclear Reactor Physics*,
https://doi.org/10.1007/978-3-031-98532-4

MIX
Papier aus verantwortungsvollen Quellen
Paper from responsible sources
FSC® C105338

If you have any concerns about our products,
you can contact us on
ProductSafety@springernature.com

In case Publisher is established outside the EU,
the EU authorized representative is:
Springer Nature Customer Service Center GmbH
Europaplatz 3, 69115 Heidelberg, Germany

Printed by Libri Plureos GmbH
in Hamburg, Germany